I0392610

# Linear Algebra

## Study Guide

Dr. Noah Ras

Copyright 2016

noah.950@live.com

# What is Linear Algebra?

Many difficult engineering problems can be handled using the powerful yet easy to use mathematics of linear algebra. Unfortunately, because the subject requires seemingly arcane and tedious computa-tions involving large arrays of number known as matrices, the key concepts and the wide applicability of linear algebra are easily missed. Therefore, before we equip you with matrix skills, let us give you some hints about what linear algebra is. The takeaway message is

> Linear algebra is the study of vectors and linear transformations.

Vectors are things you can add and linear functions are very special functions of vectors that respect vector addition. To understand this a little better, lets try some examples. Be prepared to change the way you think about some familiar mathematical objects and keep a pencil and piece of paper handy!

## What Are Vectors?

**Example 1** (Vector Addition)

(A) Numbers: If $x$ and $y$ are numbers then so is $x + y$.

(B) 3-vectors: $\begin{pmatrix} 1 \\ 1 \\ 0 \end{pmatrix} + \begin{pmatrix} 0 \\ 1 \\ 1 \end{pmatrix} = \begin{pmatrix} 1 \\ 2 \\ 1 \end{pmatrix}$.

(C) Polynomials: If $p(x) = 1 + x - 2x^2 + 3x^3$ and $q(x) = x - 3x^2 - 3x^3 - x^4$ then their sum $p(x) - q(x)$ is the new polynomial $1 + 2x + x^2 + x^4$.

(D) Power series: If $f(x) = 1 + x - \frac{1}{2!}x^2 + \frac{1}{3!}x^3 + \cdots$ and $g(x) = 1 - x - \frac{1}{2!}x^2 - \frac{1}{3!}x^3 + \cdots$ then $f(x) + g(x) = 1 - \frac{1}{2!}x^2 + \frac{1}{4!}x^4 \cdots$ is also a power series.

(E) Functions: If $f(x) = e^x$ and $g(x) = e^{-x}$ then their sum $f(x) - g(x)$ is the new function $2\cosh x$.

Stacks of numbers are not the only things that are vectors, as examples C,D, and E show. Because they "can be added", start thinking of all the above objects as vectors! In the above examples, however, notice that the vector addition rule stems from the rules for adding numbers.

When adding the same vector over and over, for example

$$x \mid x, \quad x \mid x \mid x, \quad x \mid x \mid x \mid x, \ldots,$$

we will write

$$2x, \quad 3x, \quad 4x, \ldots,$$

respectively. For example

$$4 \begin{pmatrix} 1 \\ 1 \\ 0 \end{pmatrix} = \begin{pmatrix} 1 \\ 1 \\ 0 \end{pmatrix} + \begin{pmatrix} 1 \\ 1 \\ 0 \end{pmatrix} + \begin{pmatrix} 1 \\ 1 \\ 0 \end{pmatrix} + \begin{pmatrix} 1 \\ 1 \\ 0 \end{pmatrix} = \begin{pmatrix} 4 \\ 4 \\ 0 \end{pmatrix}.$$

Defining $4x = x + x + x + x$ is fine for integer multiples, but does not help us make sense of $\frac{1}{3}x$. For the different types of vectors above, you can probably guess how to multiply a vector by a scalar. For example

$$\frac{1}{3} \begin{pmatrix} 1 \\ 1 \\ 0 \end{pmatrix} = \begin{pmatrix} \frac{1}{3} \\ \frac{1}{3} \\ 0 \end{pmatrix}.$$

In any given situation that you plan to describe using vectors, you need to decide on a way to add and scalar multiply vectors. In summary:

> Vectors are things you can add and scalar multiply.

Examples of kinds of vectors:

- numbers

- n-vectors

- 2nd order polynomials

- n-th order polynomials

- power series

- functions with a certain domain

## What Are Linear Functions?

In calculus classes, the main subject of investigation was the rates of change of functions. In linear algebra, functions will again be focus of your attention, but functions of a very special type. In calculus you were perhaps encouraged to think of a function as a machine "$f$" into which one may feed a real number. For each input $x$ this machine outputs a single real number $f(x)$.

In linear algebra, the functions we study will take vectors (of some type) as both inputs and outputs. We just saw vectors are objects that can be added or scalar multiplied—a very general notion—so the functions we are going to study will look novel at first. So things don't get too abstract, here are five questions that can be rephrased in terms of functions of vectors.

**Example 2** (Functions of Vectors in Disguise)

(A) What number $x$ solves $10x = 3$?

(B) What vector $u$ from 3-space satisfies the cross product equation $\begin{pmatrix} 1 \\ 1 \\ 0 \end{pmatrix} \times u = \begin{pmatrix} 0 \\ 1 \\ 1 \end{pmatrix}$?

(C) What polynomial $p$ satisfies $\int_{-1}^{1} p(y)dy = 0$ and $\int_{-1}^{1} yp(y)dy = 1$?

(D) What power series $f(x)$ satisfies $x\frac{d}{dx}f(x) - 2f(x) = 0$?

(E) What number $x$ solves $4x^2 = 1$?

For part (A), the machine needed would look like the picture below.

$x \longrightarrow \quad \longrightarrow 10x$,

This is just like a function $f$ from calculus that takes in a number $x$ and spits out the number $f(x) = 10x$. For part (B), we need something more sophisticated.

$$\begin{pmatrix} x \\ y \\ z \end{pmatrix} \longrightarrow \quad \longrightarrow \begin{pmatrix} z \\ -z \\ y-x \end{pmatrix},$$

The inputs and outputs are both 3-vectors. The output is the cross product of the input with... how about you complete this sentence to make sure you understand.

The machine needed for example (C) looks like it has just one input and two outputs: we input a polynomial and get a 2-vector as output.

$$p \rightarrow \blacksquare \rightarrow \begin{pmatrix} \int_{-1}^{1} p(y)dy \\ \int_{-1}^{1} yp(y)dy \end{pmatrix}.$$

This example is important because it displays an important feature; the inputs for this function are functions.

By now you may be feeling overwhelmed and thinking that absolutely any function with any kind of vector as input and any other kind of vector as output can pop up next to strain your brain! Rest assured that linear algebra involves the study of only a very simple (yet very important) class of functions of vectors; its time to describe the essential characteristics of linear functions.

Let's use the letter $L$ to denote an arbitrary linear function and think again about vector addition and scalar multiplication. Lets suppose $v$ and $u$ are vectors and $c$ is a number. Since $L$ is a function from vectors to vectors, if we input $u$ into $L$, the output $L(u)$ will also be some sort of vector. The same goes for $L(v)$. (And remember, our input and output vectors might be something other than stacks of numbers!) Because vectors are things that can be aded and scalar multiplied, $u + v$ and $cu$ are also vectors, and so they can be used as inputs. The essential characteristic of linear functions is what can be said about the outputs $L(u + v)$ and $L(cu)$.

Before we tell you this essential characteristic, ruminate on this picture.

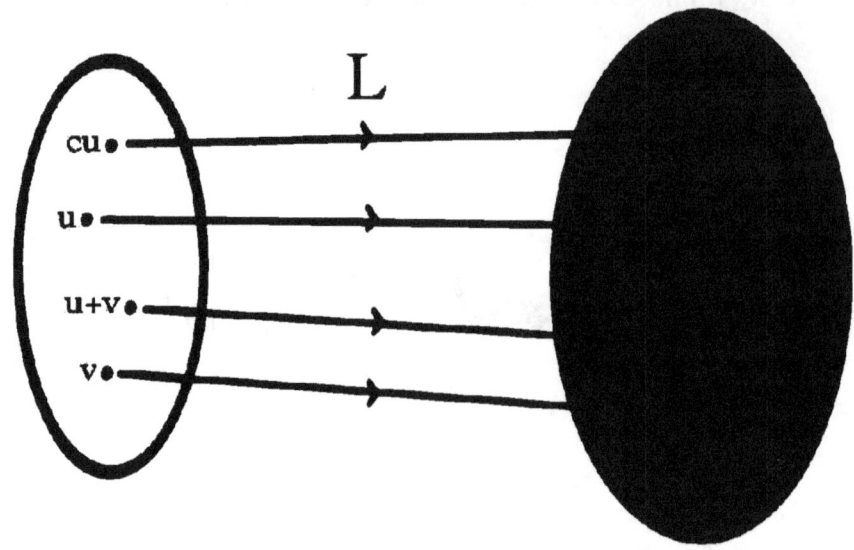

The "blob" on the left represents all the vectors that you are allowed to input into the function $L$, the blob on the right denotes the possible outputs, and the lines tell you which inputs are turned into which outputs.[1] A full pictorial description of the functions would require all inputs and outputs and lines to be explicitly drawn, but we are being diagrammatic; we only drew four of each.

## Functions have three parts

Think about adding $L(u)$ and $L(v)$ to get yet another vector $L(u) + L(v)$ or of multiplying $L(u)$ by $c$ to obtain the vector $cL(u)$, and placing both on the right blob of this picture. But wait! Are you certain that these are possible outputs!?

Here's the answer

The key to the whole class, from which everything else follows:

---

[1]The domain, codomain, and rule of correspondence of the function are represented by the left blog, right blob, and arrows, respectively.

1. Additivity:

$$L(u + v) = L(u) + L(v).$$

2. Homogeneity:

$$L(cu) = cL(u).$$

Most functions of vectors do not obey this requirement.[2] At its heart, linear algebra is the study of functions that do.

Notice that the additivity requirement says that the function $L$ respects vector addition: *it does not matter if you first add u and v and then input their sum into L, or first input u and v into L separately and then add the outputs*. The same holds for scalar multiplication–try writing out the scalar multiplication version of the italicized sentence. When a function of vectors obeys the additivity and homogeneity properties we say that it is *linear* (this is the "linear" of linear algebra). Together, additivity and homogeneity are called *linearity*. Are there other, equivalent, names for linear functions? yes.

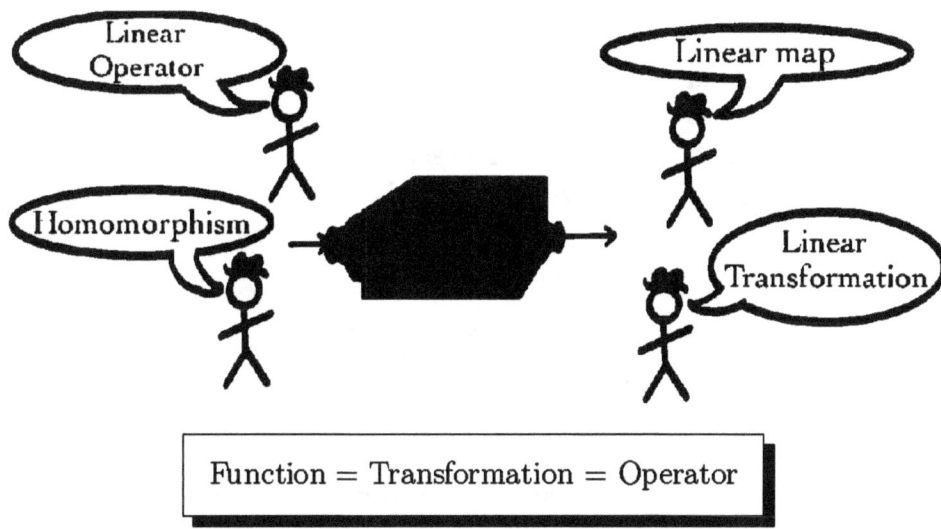

$$\boxed{\text{Function} = \text{Transformation} = \text{Operator}}$$

And now for a hint at the power of linear algebra. The questions in examples (A-D) can all be restated as

$$\boxed{Lv = w}$$

---

[2]e.g. If $f(x) = x^2$ then $f(1 + 1) = 4 \neq f(1) + f(1) = 2$. Try any other function you can think of!

where $v$ is an unknown, $w$ a known vector, and $L$ is a known linear transformation. To check that this is true, one needs to know the rules for adding vectors (both inputs and outputs) and then check linearity of $L$. Solving the equation $Lv = w$ often amounts to solving systems of linear equations, the skill you will learn in Chapter 2.

A great example is the derivative operator.

**Example 3** (The derivative operator is linear)
For any two functions $f(x)$, $g(x)$ and any number $c$, in calculus you probably learnt that the derivative operator satisfies

1. $\frac{d}{dx}(cf) = c\frac{d}{dx}f$,

2. $\frac{d}{dx}(f+g) = \frac{d}{dx}f + \frac{d}{dx}g$.

If we view functions as vectors with addition given by addition of functions and with scalar multiplication given by multiplication of functions by constants, then these familiar properties of derivatives are just the linearity property of linear maps.

Before introducing matrices, notice that for linear maps $L$ we will often write simply $Lu$ instead of $L(u)$. This is because the linearity property of a linear transformation $L$ means that $L(u)$ can be thought of as multiplying the vector $u$ by the linear operator $L$. For example, the linearity of $L$ implies that if $u, v$ are vectors and $c, d$ are numbers, then

$$L(cu + dv) = cLu + dLv \,,$$

which feels a lot like the regular rules of algebra for numbers. Notice though, that "$uL$" makes no sense here.

**Remark** A sum of multiples of vectors $cu + dv$ is called a *linear combination* of $u$ and $v$.

## What is a Matrix?

Matrices are linear functions of a certain kind. One way to learn about them is by studying *systems of linear equations*.

**Example 4** A room contains $x$ bags and $y$ boxes of fruit:

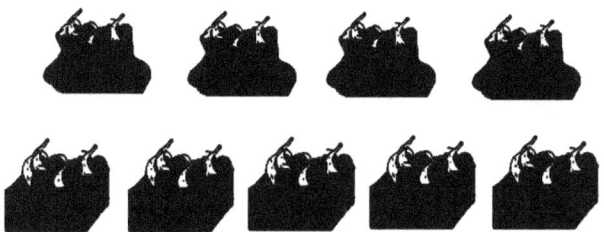

Each bag contains 2 apples and 4 bananas and each box contains 6 apples and 8 bananas. There are 20 apples and 28 bananas in the room. Find $x$ and $y$.

The values are the numbers $x$ and $y$ that simultaneously make both of the following equations true:

$$2x + 6y = 20$$
$$4x + 8y = 28.$$

Here we have an example of a *System of Linear Equations*.[3] It's a collection of equations in which variables are multiplied by constants and summed, and no variables are multiplied together: There are no powers of variables (like $x^2$ or $y^5$), non-integer or negative powers of variables (like $y^{1/7}$ or $x^{-3}$), and no places where variables are multiplied together (like $xy$).

Information about the fruity contents of the room can be stored two ways:

(i) In terms of the number of apples and bananas.

(ii) In terms of the number of bags and boxes.

Intuitively, knowing the information in one form allows you to figure out the information in the other form. Going from (ii) to (i) is easy: If you knew there were 3 bags and 2 boxes it would be easy to calculate the number of apples and bananas, and doing so would have the feel of multiplication

---

[3] Perhaps you can see that both lines are of the form $Lu = v$ with $u \begin{pmatrix} x \\ y \end{pmatrix}$ an unknown, $v = 20$ in the first line, $v = 28$ in the second line, and $L$ different functions in each line? We give the typical less sophisticated description in the text above.

(containers times fruit per container). In the example above we are required to go the other direction, from (i) to (ii). This feels like the opposite of multiplication, *i.e.*, division. Matrix notation will make clear what we are "multiplying" and "dividing" by.

The goal of Chapter 2 is to efficiently solve systems of linear equations. Partly, this is just a matter of finding a better notation, but one that hints at a deeper underlying mathematical structure. For that, we need rules for adding and scalar multiplying 2-vectors:

$$c \begin{pmatrix} x \\ y \end{pmatrix} := \begin{pmatrix} cx \\ cy \end{pmatrix} \quad \text{and} \quad \begin{pmatrix} x \\ y \end{pmatrix} + \begin{pmatrix} x' \\ y' \end{pmatrix} := \begin{pmatrix} x + x' \\ y + y' \end{pmatrix}.$$

Writing our fruity equations as an equality between 2-vectors and then using these rules we have:

$$\left. \begin{array}{c} 2x + 6y = 20 \\ 4x + 8y = 28 \end{array} \right\} \iff \begin{pmatrix} 2x + 6y \\ 4x + 8y \end{pmatrix} = \begin{pmatrix} 20 \\ 28 \end{pmatrix} \iff x \begin{pmatrix} 2 \\ 4 \end{pmatrix} + y \begin{pmatrix} 6 \\ 8 \end{pmatrix} = \begin{pmatrix} 20 \\ 28 \end{pmatrix}.$$

Now we introduce an function which takes in 2-vectors[4] and gives out 2-vectors. We denote it by an array of numbers called a *matrix* .

**The function** $\begin{pmatrix} 2 & 6 \\ 4 & 8 \end{pmatrix}$ **is defined by** $\begin{pmatrix} 2 & 6 \\ 4 & 8 \end{pmatrix} \begin{pmatrix} x \\ y \end{pmatrix} := x \begin{pmatrix} 2 \\ 4 \end{pmatrix} + y \begin{pmatrix} 6 \\ 8 \end{pmatrix}.$

A similar definition applies to matrices with different numbers and sizes.

**Example 5** (A bigger matrix)

$$\begin{pmatrix} 1 & 0 & 3 & 4 \\ 5 & 0 & 3 & 4 \\ -1 & 6 & 2 & 5 \end{pmatrix} \begin{pmatrix} x \\ y \\ z \\ w \end{pmatrix} := x \begin{pmatrix} 1 \\ 5 \\ -1 \end{pmatrix} - y \begin{pmatrix} 0 \\ 0 \\ 6 \end{pmatrix} + z \begin{pmatrix} 3 \\ 3 \\ 2 \end{pmatrix} + w \begin{pmatrix} 4 \\ 4 \\ 5 \end{pmatrix}.$$

---

[4]To be clear, we will use the term 2-vector to refer to stacks of two numbers such as $\begin{pmatrix} 7 \\ 11 \end{pmatrix}$. If we wanted to refer to the vectors $x^2 + 1$ and $x^3 - 1$ (recall that polynomials are vectors) we would say "consider the two vectors $x^3 - 1$ and $x^2 + 1$. We apologize through giggles for the possibility of the phrase "two 2-vectors."

Viewed as a machine that inputs and outputs 2-vectors, our $2 \times 2$ matrix does the following:

$$\begin{pmatrix} x \\ y \end{pmatrix} \longrightarrow \boxed{\phantom{xxx}} \longrightarrow \begin{pmatrix} 2x + 6y \\ 4x + 8y \end{pmatrix}.$$

Our fruity problem is now rather concise.

**Example 6** (This time in purely mathematical language):

What vector $\begin{pmatrix} x \\ y \end{pmatrix}$ satisfies $\begin{pmatrix} 2 & 6 \\ 4 & 8 \end{pmatrix} \begin{pmatrix} x \\ y \end{pmatrix} = \begin{pmatrix} 20 \\ 28 \end{pmatrix}$?

This is of the same $Lv = w$ form as our opening examples. The matrix encodes fruit per container. The equation is roughly fruit per container times number of containers equals fruit. To solve for number of containers we want to somehow "divide" by the matrix.

Another way to think about the above example is to remember the rule for multiplying a matrix times a vector. If you have forgotten this, you can actually guess a good rule by making sure the matrix equation is the same as the system of linear equations. This would require that

$$\begin{pmatrix} 2 & 6 \\ 4 & 8 \end{pmatrix} \begin{pmatrix} x \\ y \end{pmatrix} := \begin{pmatrix} 2x + 6y \\ 4x + 8y \end{pmatrix}$$

Indeed this is an example of the general rule that you have probably seen before

$$\begin{pmatrix} p & q \\ r & s \end{pmatrix} \begin{pmatrix} x \\ y \end{pmatrix} := \begin{pmatrix} px + qy \\ rx + sy \end{pmatrix} = x \begin{pmatrix} p \\ r \end{pmatrix} + y \begin{pmatrix} q \\ s \end{pmatrix}.$$

Notice, that the second way of writing the output on the right hand side of this equation is very useful because it tells us what all possible outputs a matrix times a vector look like – they are just sums of the columns of the matrix multiplied by scalars. The set of all possible outputs of a matrix times a vector is called the **column space** (it is also the image of the linear function defined by the matrix).

A matrix is an example of a *Linear Function*, because it takes one vector and turns it into another in a "linear" way. Of course, we can have much larger matrices if our system has more variables.

## Matrices in Space!

Matrices are linear functions. The statement of this for the matrix in our fruity example is as follows.

1. $\begin{pmatrix} 2 & 6 \\ 4 & 8 \end{pmatrix} c \begin{pmatrix} x \\ y \end{pmatrix} = c \begin{pmatrix} 2 & 6 \\ 4 & 8 \end{pmatrix} \begin{pmatrix} a \\ b \end{pmatrix}$ and

2. $\begin{pmatrix} 2 & 6 \\ 4 & 8 \end{pmatrix} \left[ \begin{pmatrix} x \\ y \end{pmatrix} + \begin{pmatrix} x' \\ y' \end{pmatrix} \right] = \begin{pmatrix} 2 & 6 \\ 4 & 8 \end{pmatrix} \begin{pmatrix} x \\ y \end{pmatrix} + \begin{pmatrix} 2 & 6 \\ 4 & 8 \end{pmatrix} \begin{pmatrix} x' \\ y' \end{pmatrix}.$

These equalities can be verified using the rules we introduced so far.

**Example 7** Verify that $\begin{pmatrix} 2 & 6 \\ 4 & 8 \end{pmatrix}$ is a linear operator.

The matrix is homogeneous if the expressions on the left hand side and right hand side of the first equation are indeed equal.

$$\begin{pmatrix} 2 & 6 \\ 4 & 8 \end{pmatrix} \left[ c \begin{pmatrix} a \\ b \end{pmatrix} \right] = \begin{pmatrix} 2 & 6 \\ 4 & 8 \end{pmatrix} \begin{pmatrix} ca \\ cb \end{pmatrix} = ca \begin{pmatrix} 2 \\ 4 \end{pmatrix} + cb \begin{pmatrix} 6 \\ 8 \end{pmatrix}$$

$$= \begin{pmatrix} 2ac \\ 4ac \end{pmatrix} + \begin{pmatrix} 6bc \\ 8bc \end{pmatrix} = \underline{\begin{pmatrix} 2ac + 6bc \\ 4ac + 8bc \end{pmatrix}}$$

while

$$c \begin{pmatrix} 2 & 6 \\ 4 & 8 \end{pmatrix} \begin{pmatrix} a \\ b \end{pmatrix} = c \left[ a \begin{pmatrix} 2 \\ 4 \end{pmatrix} + b \begin{pmatrix} 6 \\ 8 \end{pmatrix} \right] = c \left[ \begin{pmatrix} 2a \\ 4a \end{pmatrix} + \begin{pmatrix} 6b \\ 8b \end{pmatrix} \right]$$

$$= c \begin{pmatrix} 2a + 6b \\ 4a + 8b \end{pmatrix} = \underline{\begin{pmatrix} 2ac + 6bc \\ 4ac + 8bc \end{pmatrix}}.$$

The underlined expressions are visually identical, so the matrix is homogeneous.

The matrix is additive if the left and right side of the second equation are indeed equal.

$$\begin{pmatrix} 2 & 6 \\ 4 & 8 \end{pmatrix} \left[ \begin{pmatrix} a \\ b \end{pmatrix} + \begin{pmatrix} c \\ d \end{pmatrix} \right] = \begin{pmatrix} 2 & 6 \\ 4 & 8 \end{pmatrix} \begin{pmatrix} a+c \\ b+d \end{pmatrix} = (a+c) \begin{pmatrix} 2 \\ 4 \end{pmatrix} + (b+d) \begin{pmatrix} 6 \\ 8 \end{pmatrix}$$

$$= \begin{pmatrix} 2(a+c) \\ 4(a+c) \end{pmatrix} + \begin{pmatrix} 6(b+d) \\ 8(b+d) \end{pmatrix} = \begin{pmatrix} 2a+2c+6b+6d \\ 4a+4c+8b+8d \end{pmatrix}$$

which we need to compare to

$$\begin{pmatrix} 2 & 6 \\ 4 & 8 \end{pmatrix} \begin{pmatrix} a \\ b \end{pmatrix} + \begin{pmatrix} 2 & 6 \\ 4 & 8 \end{pmatrix} \begin{pmatrix} c \\ d \end{pmatrix} = a \begin{pmatrix} 2 \\ 4 \end{pmatrix} + b \begin{pmatrix} 6 \\ 8 \end{pmatrix} + c \begin{pmatrix} 2 \\ 4 \end{pmatrix} + d \begin{pmatrix} 6 \\ 8 \end{pmatrix}$$

$$= \begin{pmatrix} 2a \\ 4a \end{pmatrix} + \begin{pmatrix} 6b \\ 8b \end{pmatrix} + \begin{pmatrix} 2c \\ 4c \end{pmatrix} + \begin{pmatrix} 6d \\ 8d \end{pmatrix} = \begin{pmatrix} 2a+2c+6b+6d \\ 4a+4c+8b+8d \end{pmatrix}.$$

The matrix is additive and homogeneous, and so it is, by definition, linear.

We have come full circle; matrices are just examples of the kinds of linear operators that appear in algebra problems like those in section 1.2. Any equation of the form $Mv = w$ with $M$ a matrix, and $v, w$ $n$-vectors is called a *matrix equation*. Chapter 2 is about efficiently solving systems of linear equations, or equivalently matrix equations.

## The Matrix Detour

Linear algebra is about linear functions, not matrices. This lesson is hard to learn after a full term of working with matrices so we want to get you thinking about this on day one of the course. We hope you will be thinking about this idea constantly throughout the course.

> Matrices only get involved in linear algebra when certain notational choices are made.

To exemplify, lets look at the derivative operator again.

**Example 8** of how matrices come into linear algebra.
Consider the equation

$$\left( \frac{d}{dx} + 2 \right) f = x + 1$$

where $f$ is unknown (the place where solutions should go) and the linear differential operator $\frac{d}{dx} - 2$ is understood to take in quadratic functions (of the form $ax^2 - bx - c$) and give out other quadratic functions.

Let's simplify the way we denote the quadratic functions; we will

$$\text{denote } ax^2 - bx + c \text{ as } \begin{pmatrix} a \\ b \\ c \end{pmatrix}_B .$$

The subscript $B$ serves to remind us of our particular notional convention; we will compare to another notational convention later. With the convention $B$ we can say

$$\left( \frac{d}{dx} - 2 \right) \begin{pmatrix} a \\ b \\ c \end{pmatrix}_B = \left( \frac{d}{dx} + 2 \right) (ax^2 + bx - c)$$

$$= (2ax - b) + (2ax^2 - 2bx - 2c) = 2ax^2 + (2a + 2b)x - (b + 2c)$$

$$= \begin{pmatrix} 2a \\ 2a + 2b \\ b + 2c \end{pmatrix}_B = \left[ \begin{pmatrix} 2 & 0 & 0 \\ 2 & 2 & 0 \\ 0 & 1 & 2 \end{pmatrix} \begin{pmatrix} a \\ b \\ c \end{pmatrix} \right]_B .$$

That is, our notational convention for quadratic functions has induced a notation for the differential operator $\frac{d}{dx} + 2$ as a matrix. We can use this notation to change the way that the following two equations say exactly the same thing.

$$\left( \frac{d}{dx} + 2 \right) f = x - 1 \Leftrightarrow \left[ \begin{pmatrix} 2 & 0 & 0 \\ 2 & 2 & 0 \\ 0 & 1 & 2 \end{pmatrix} \begin{pmatrix} a \\ b \\ c \end{pmatrix} \right]_B = \begin{pmatrix} 0 \\ 1 \\ 1 \end{pmatrix}_B .$$

Our notational convention has served as an organizing principle to yield the system of equations

$$\begin{aligned} 2a &= 0 \\ 2a + 2b &= 1 \\ b - 2c &= 1 \end{aligned}$$

with solution $\begin{pmatrix} 0 \\ \frac{1}{2} \\ \frac{1}{4} \end{pmatrix}$ which the notational convention $B$ uses to encode $\frac{1}{2}x + \frac{1}{4}$, which is indeed the solution to our equation since, substituting for $f$ yields the true statement $\left( \frac{d}{dx} - 2 \right) \left( \frac{1}{2}x - \frac{1}{4} \right) = x - 1$.

It would be nice to have a systematic way to rewrite any linear equation as an equivalent matrix equation. It will be a while before we can make rigorous this way of organizing information in a way generalizable to all linear equations, but keep this example in mind throughout the course.

The general idea is presented in the picture below; sometimes a linear equation is too hard to solve as is, but by reformulating it into a matrix equation the process of finding solutions becomes doable.

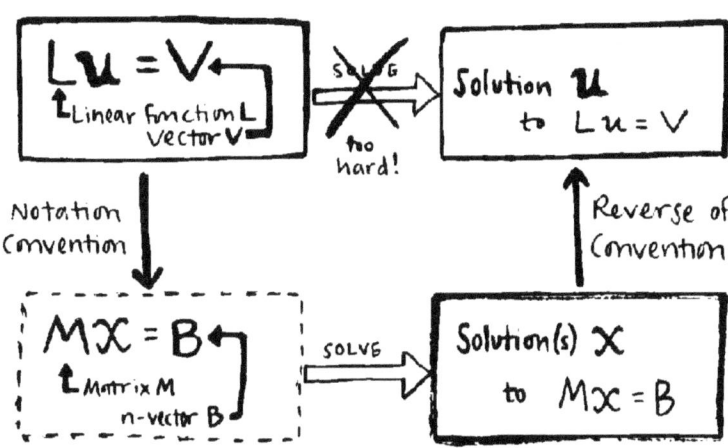

A simple example with the knowns ($L$ and $V$ are $\frac{d}{dx}$ and 3, respectively) is shown below, although the detour is unnecessary in this case since you know how to anti-differentiate.

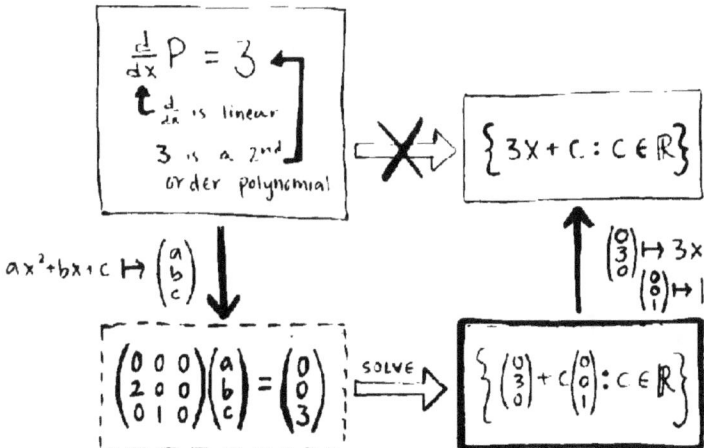

To drive home the point that we are not studying matrices but rather linear functions, and that those linear functions can be represented as matrices under certain notational conventions, consider how changeable the notational conventions are.

**Example 9** of how a different matrix comes into the same linear algebra problem.

Another possible notational convention is to

$$\text{denote } a - bx + cx^2 \text{ as } \begin{pmatrix} a \\ b \\ c \end{pmatrix}_{B'}.$$

With this alternative notation

$$\left(\frac{d}{dx} - 2\right) \begin{pmatrix} a \\ b \\ c \end{pmatrix}_{B'} = \left(\frac{d}{dx} - 2\right)(a + bx - cx^2)$$

$$= (b + 2cx) + (2a + 2bx - 2cx^2) = (2a + b) - (2b + 2c)x - 2cx^2$$

$$= \begin{pmatrix} 2a + b \\ 2b - 2c \\ 2c \end{pmatrix}_{B'} = \left[\begin{pmatrix} 2 & 1 & 0 \\ 0 & 2 & 2 \\ 0 & 0 & 2 \end{pmatrix} \begin{pmatrix} a \\ b \\ c \end{pmatrix}\right]_{B'}.$$

Notice that we have obtained *a different matrix for the same linear function*. The equation we started with

$$\left(\frac{d}{dx} - 2\right)f = x + 1 \Leftrightarrow \left[\begin{pmatrix} 2 & 1 & 0 \\ 0 & 2 & 2 \\ 0 & 0 & 2 \end{pmatrix} \begin{pmatrix} a \\ b \\ c \end{pmatrix}\right]_{B'} = \begin{pmatrix} 1 \\ 1 \\ 0 \end{pmatrix}_{B'}$$

$$\Leftrightarrow \begin{array}{r} 2a + b = 1 \\ 2b - 2c = 1 \\ 2c = 0 \end{array}$$

has the solution $\begin{pmatrix} \frac{1}{4} \\ \frac{1}{2} \\ 0 \end{pmatrix}$. Notice that we have obtained *a different 3-vector for the same vector*, since in the notational convention $B'$ this 3-vector represents $\frac{1}{4} + \frac{1}{2}x$.

One linear function can be represented (denoted) by a huge variety of matrices. The representation only depends on how vectors are denoted as n-vectors.

## Review Problems

You probably have already noticed that understanding sets, functions and basic logical operations is a must to do well in linear algebra. Brush up on these skills by trying these background webwork problems:

| | |
|---|---|
| Logic | 1 |
| Sets | 2 |
| Functions | 3 |
| Equivalence Relations | 4 |
| Proofs | 5 |

Each chapter also has reading and skills WeBWorK problems:

**Webwork:** | Reading problems | 1 , 2 |

Probably you will spend most of your time on the following review questions.

1. Problems A, B, and C of example 2 can all be written as $Lv = w$ where

$$L : V \longrightarrow W,$$

(read this as $L$ maps the set of vectors $V$ to the set of vectors $W$). For each case write down the sets $V$ and $W$ where the vectors $v$ and $w$ come from.

2. Torque is a measure of "rotational force". It is a vector whose direction is the (preferred) axis of rotation. Upon applying a force $F$ on an object at point $r$ the torque $\tau$ is the cross product $r \times F = \tau$.

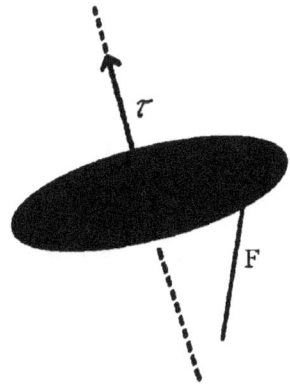

Lets find the force $F$ (a vector) one must apply to a wrench lying along the vector $r = \begin{pmatrix} 1 \\ 1 \\ 0 \end{pmatrix}$ ft, to produce a torque $\begin{pmatrix} 0 \\ 0 \\ 1 \end{pmatrix}$ ft lb:

(a) Find a solution by writing out this equation with $F = \begin{pmatrix} a \\ b \\ c \end{pmatrix}$.

(Hint: Guess and check that a solution with $a = 0$ exists).

(b) Add $\begin{pmatrix} 1 \\ 1 \\ 0 \end{pmatrix}$ to your solution and check that the result is a solution.

(c) Give a physics explanation of why there can be two solutions, and argue that there are, in fact, infinitely many solutions.

(d) Set up a system of three linear equations with the three components of $F$ as the variables which describes this situation. What happens if you try to solve these equations by substitution?

3. The function $P(t)$ gives gas prices (in units of dollars per gallon) as a function of $t$ the year (in A.D. or C.E.), and $g(t)$ is the gas consumption rate measured in gallons per year by a driver as a function of their age. The function $g$ is certainly different for different people. Assuming a lifetime is 100 years, what function gives the total amount spent on gas during the lifetime of an individual born in an arbitrary year $t$? Is the operator that maps $g$ to this function linear?

4. The differential equation (DE)

$$\frac{d}{dt}f = 2f$$

says that the rate of change of $f$ is proportional to $f$. It describes exponential growth because the exponential function

$$f(t) = f(0)e^{2t}$$

satisfies the DE for any number $f(0)$. The number 2 in the DE is called the constant of proportionality. A similar DE

$$\frac{d}{dt}f = \frac{2}{t}f$$

has a time-dependent "constant of proportionality".

(a) Do you think that the second DE describes exponential growth?

(b) Write both DEs in the form $Df = 0$ with $D$ a linear operator.

5. Pablo is a nutritionist who knows that oranges always have twice as much sugar as apples. When considering the sugar intake of schoolchildren eating a barrel of fruit, he represents the barrel like so:

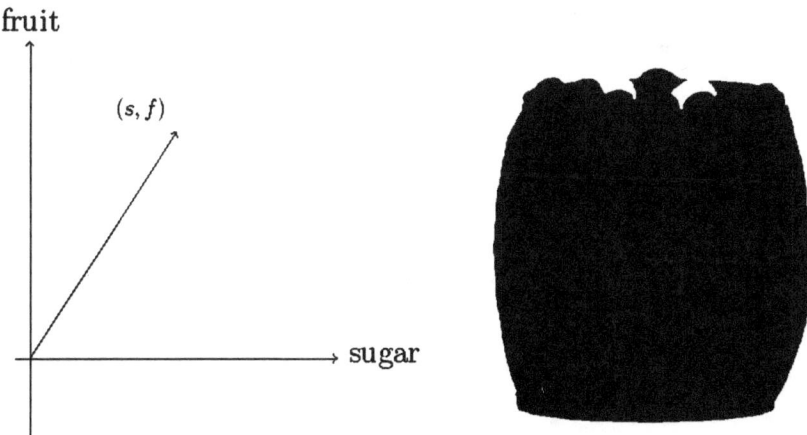

Find a linear operator relating Pablo's representation to the "everyday" representation in terms of the number of apples and number of oranges. Write your answer as a matrix.

Hint: Let $\lambda$ represent the amount of sugar in each apple.

## Hint

6. *Matrix Multiplication:* Let $M$ and $N$ be matrices

$$M = \begin{pmatrix} a & b \\ c & d \end{pmatrix} \text{ and } N = \begin{pmatrix} e & f \\ g & h \end{pmatrix},$$

and $v$ the vector

$$v = \begin{pmatrix} x \\ y \end{pmatrix}.$$

If we first apply $N$ and then $M$ to $v$ we obtain the vector $MNv$.

(a) Show that the composition of matrices $MN$ is also a linear operator.

(b) Write out the components of the matrix product $MN$ in terms of the components of $M$ and the components of $N$. *Hint*: use the general rule for multiplying a 2-vector by a 2×2 matrix.

(c) Try to answer the following common question, "Is there any sense in which these rules for matrix multiplication are unavoidable, or are they just a notation that could be replaced by some other notation?"

(d) Generalize your multiplication rule to $3 \times 3$ matrices.

7. *Diagonal matrices:* A matrix $M$ can be thought of as an array of numbers $m^i_j$, known as matrix entries, or matrix components, where $i$ and $j$ index row and column numbers, respectively. Let

$$M = \begin{pmatrix} 1 & 2 \\ 3 & 4 \end{pmatrix} = (m^i_j).$$

Compute $m^1_1$, $m^1_2$, $m^2_1$ and $m^2_2$.

The matrix entries $m^i_i$ whose row and column numbers are the same are called the *diagonal* of $M$. Matrix entries $m^i_j$ with $i \neq j$ are called *off-diagonal*. How many diagonal entries does an $n \times n$ matrix have? How many off-diagonal entries does an $n \times n$ matrix have?

If all the off-diagonal entries of a matrix vanish, we say that the matrix is diagonal. Let

$$D = \begin{pmatrix} \lambda & 0 \\ 0 & \mu \end{pmatrix} \quad \text{and} \quad D' = \begin{pmatrix} \lambda' & 0 \\ 0 & \mu' \end{pmatrix}.$$

Are these matrices diagonal and why? Use the rule you found in problem 6 to compute the matrix products $DD'$ and $D'D$. What do you observe? Do you think the same property holds for arbitrary matrices? What about products where only one of the matrices is diagonal?

(p.s. Diagonal matrices play a special role in in the study of matrices in linear algebra. Keep an eye out for this special role.)

8. Find the linear operator that takes in vectors from $n$-space and gives out vectors from $n$-space in such a way that

   (a) whatever you put in, you get exactly the same thing out as what you put in. Show that it is unique. Can you write this operator as a matrix?

   (b) whatever you put in, you get exactly the same thing out as when you put something else in. Show that it is unique. Can you write this operator as a matrix?

   *Hint:* To show something is unique, it is usually best to begin by pretending that it isn't, and then showing that this leads to a nonsensical conclusion. In mathspeak–*proof by contradiction*.

9. Consider the set $S = \{*, \star, \#\}$. It contains just 3 elements, and has no ordering; $\{*, \star, \#\} = \{\#, \star, *\}$ etc. (In fact the same is true for $\{1, 2, 3\} = \{2, 3, 1\}$ etc, although we could make this an *ordered set* using $3 > 2 > 1$.)

   (i) Invent a function with domain $\{*, \star, \#\}$ and codomain $\mathbb{R}$. (Remember that the *domain* of a function is the set of all its allowed inputs and the *codomain* (or *target space*) is the set where the outputs can live. A function is specified by assigning exactly one codomain element to each element of the domain.)

   (ii) Choose an ordering on $\{*, \star, \#\}$, and then use it to write your function from part (i) as a triple of numbers.

   (iii) Choose a new ordering on $\{*, \star, \#\}$ and then write your function from part (i) as a triple of numbers.

   (iv) Your answers for parts (ii) and (iii) are different yet represent the same function – explain!

# Systems of Linear Equations

## Gaussian Elimination

Systems of linear equations can be written as matrix equations. Now you will learn an efficient algorithm for (maximally) simplifying a system of linear equations (or a matrix equation) – Gaussian elimination.

## Augmented Matrix Notation

Efficiency demands a new notation, called an *augmented matrix*, which we introduce via examples:

The linear system

$$\begin{cases} x + y = 27 \\ 2x - y = 0, \end{cases}$$

is denoted by the augmented matrix

$$\left( \begin{array}{cc|c} 1 & 1 & 27 \\ 2 & -1 & 0 \end{array} \right).$$

This notation is simpler than the matrix one,

$$\begin{pmatrix} 1 & 1 \\ 2 & -1 \end{pmatrix} \begin{pmatrix} x \\ y \end{pmatrix} = \begin{pmatrix} 27 \\ 0 \end{pmatrix},$$

although all three of the above denote the same thing.

## Augmented Matrix Notation

Another interesting rewriting is

$$x \begin{pmatrix} 1 \\ 2 \end{pmatrix} + y \begin{pmatrix} 1 \\ -1 \end{pmatrix} = \begin{pmatrix} 27 \\ 0 \end{pmatrix} .$$

This tells us that we are trying to find the combination of the vectors $\begin{pmatrix} 1 \\ 2 \end{pmatrix}$ and $\begin{pmatrix} 1 \\ -1 \end{pmatrix}$ adds up to $\begin{pmatrix} 27 \\ 0 \end{pmatrix}$; the answer is "clearly" $9 \begin{pmatrix} 1 \\ 2 \end{pmatrix} + 18 \begin{pmatrix} 1 \\ -1 \end{pmatrix}$.

Here is a larger example. The system

$$
\begin{aligned}
1x + 3y + 2z + 0w &= 9 \\
6x + 2y + 0z - 2w &= 0 \\
-1x + 0y + 1z + 1w &= 3 ,
\end{aligned}
$$

is denoted by the augmented matrix

$$\left( \begin{array}{cccc|c} 1 & 3 & 2 & 0 & 9 \\ 6 & 2 & 0 & -2 & 0 \\ -1 & 0 & 1 & 1 & 3 \end{array} \right) ,$$

which is equivalent to the matrix equation

$$\begin{pmatrix} 1 & 3 & 2 & 0 \\ 6 & 2 & 0 & -2 \\ -1 & 0 & 1 & 1 \end{pmatrix} \begin{pmatrix} x \\ y \\ z \\ w \end{pmatrix} = \begin{pmatrix} 9 \\ 0 \\ 3 \end{pmatrix} .$$

Again, we are trying to find which combination of the columns of the matrix adds up to the vector on the right hand side.

For the the general case of $r$ linear equations in $k$ unknowns, the number of equations is the number of rows $r$ in the augmented matrix, and the number of columns $k$ in the matrix left of the vertical line is the number of unknowns, giving an augmented matrix of the form

$$\left( \begin{array}{cccc|c} a_1^1 & a_2^1 & \cdots & a_k^1 & b^1 \\ a_1^2 & a_2^2 & \cdots & a_k^2 & b^2 \\ \vdots & \vdots & & \vdots & \vdots \\ a_1^r & a_2^r & \cdots & a_k^r & b^r \end{array} \right) .$$

Entries left of the divide carry two indices; subscripts denote column number and superscripts row number. We emphasize, the superscripts here do *not* denote exponents. Make sure you can write out the system of equations and the associated matrix equation for any augmented matrix.

We now have three ways of writing the same question. Let's put them side by side as we solve the system by strategically adding and subtracting equations. We will not tell you the motivation for this particular series of steps yet, but let you develop some intuition first.

**Example 10** (How matrix equations and augmented matrices change in elimination)

$$\left. \begin{array}{rcrcl} x & + & y & = & 27 \\ 2x & - & y & = & 0 \end{array} \right\} \Leftrightarrow \begin{pmatrix} 1 & 1 \\ 2 & -1 \end{pmatrix} \begin{pmatrix} x \\ y \end{pmatrix} = \begin{pmatrix} 27 \\ 0 \end{pmatrix} \Leftrightarrow \left( \begin{array}{cc|c} 1 & 1 & 27 \\ 2 & -1 & 0 \end{array} \right).$$

With the first equation replaced by the sum of the two equations this becomes

$$\left. \begin{array}{rcrcl} 3x & + & 0 & = & 27 \\ 2x & - & y & = & 0 \end{array} \right\} \Leftrightarrow \begin{pmatrix} 3 & 0 \\ 2 & -1 \end{pmatrix} \begin{pmatrix} x \\ y \end{pmatrix} = \begin{pmatrix} 27 \\ 0 \end{pmatrix} \Leftrightarrow \left( \begin{array}{cc|c} 3 & 0 & 27 \\ 2 & -1 & 0 \end{array} \right).$$

Let the new first equation be the old first equation divided by 3:

$$\left. \begin{array}{rcrcl} x & + & 0 & = & 9 \\ 2x & - & y & = & 0 \end{array} \right\} \Leftrightarrow \begin{pmatrix} 1 & 0 \\ 2 & -1 \end{pmatrix} \begin{pmatrix} x \\ y \end{pmatrix} = \begin{pmatrix} 9 \\ 0 \end{pmatrix} \Leftrightarrow \left( \begin{array}{cc|c} 1 & 0 & 9 \\ 2 & -1 & 0 \end{array} \right).$$

Replace the second equation by the second equation minus two times the first equation:

$$\left. \begin{array}{rcrcl} x & + & 0 & = & 9 \\ 0 & - & y & = & -18 \end{array} \right\} \Leftrightarrow \begin{pmatrix} 1 & 0 \\ 0 & -1 \end{pmatrix} \begin{pmatrix} x \\ y \end{pmatrix} = \begin{pmatrix} 9 \\ -18 \end{pmatrix} \Leftrightarrow \left( \begin{array}{cc|c} 1 & 0 & 9 \\ 0 & -1 & -18 \end{array} \right).$$

Let the new second equation be the old second equation divided by -1:

$$\left. \begin{array}{rcrcl} x & + & 0 & = & 9 \\ 0 & + & y & = & 18 \end{array} \right\} \Leftrightarrow \begin{pmatrix} 1 & 0 \\ 0 & 1 \end{pmatrix} \begin{pmatrix} x \\ y \end{pmatrix} = \begin{pmatrix} 9 \\ 18 \end{pmatrix} \Leftrightarrow \left( \begin{array}{cc|c} 1 & 0 & 9 \\ 0 & 1 & 18 \end{array} \right).$$

Did you see what the strategy was? To *eliminate y* from the first equation and then *eliminate x* from the second. The result was the solution to the system.

Here is the big idea: Everywhere in the instructions above we can replace the word "equation" with the word "row" and interpret them as telling us what to do with the augmented matrix instead of the system of equations. Performed systemically, the result is the **Gaussian elimination** algorithm.

## Equivalence and the Act of Solving

We introduce the symbol $\sim$ which is called "tilde" but should be read as "is (row) equivalent to" because at each step the augmented matrix changes by an operation on its rows but its solutions do not. For example, we found above that

$$\begin{pmatrix} 1 & 1 & | & 27 \\ 2 & -1 & | & 0 \end{pmatrix} \sim \begin{pmatrix} 1 & 0 & | & 9 \\ 2 & -1 & | & 0 \end{pmatrix} \sim \begin{pmatrix} 1 & 0 & | & 9 \\ 0 & 1 & | & 18 \end{pmatrix}.$$

The last of these augmented matrices is our favorite!

## Equivalence Example

Setting up a string of equivalences like this is a means of solving a system of linear equations. This is the main idea of section 2.1.3. This next example hints at the main trick:

**Example 11** (Using Gaussian elimination to solve a system of linear equations)

$$\left. \begin{matrix} x + y & = & 5 \\ x + 2y & = & 8 \end{matrix} \right\} \Leftrightarrow \begin{pmatrix} 1 & 1 & | & 5 \\ 1 & 2 & | & 8 \end{pmatrix} \sim \begin{pmatrix} 1 & 1 & | & 5 \\ 0 & 1 & | & 3 \end{pmatrix} \sim \begin{pmatrix} 1 & 0 & | & 2 \\ 0 & 1 & | & 3 \end{pmatrix} \Leftrightarrow \left\{ \begin{matrix} x + 0 & = & 2 \\ 0 + y & = & 3 \end{matrix} \right.$$

Note that in going from the first to second augmented matrix, we used the top left 1 to make the bottom left entry zero. For this reason we call the top left entry a pivot. Similarly, to get from the second to third augmented matrix, the bottom right entry (before the divide) was used to make the top right one vanish; so the bottom right entry is also called a pivot.

This name *pivot* is used to indicate the matrix entry used to "zero out" the other entries in its column; the pivot is the number used to eliminate another number in its column.

## Reduced Row Echelon Form

For a system of two linear equations, the goal of Gaussian elimination is to convert the part of the augmented matrix left of the dividing line into the matrix

$$I = \begin{pmatrix} 1 & 0 \\ 0 & 1 \end{pmatrix},$$

called the *Identity Matrix*, since this would give the simple statement of a solution $x = a, y = b$. The same goes for larger systems of equations for which the identity matrix $I$ has 1's along its diagonal and all off-diagonal entries vanish:

$$I = \begin{pmatrix} 1 & 0 & \cdots & 0 \\ 0 & 1 & & 0 \\ \vdots & & \ddots & \vdots \\ 0 & 0 & \cdots & 1 \end{pmatrix}$$

For many systems, it is not possible to reach the identity in the augmented matrix via Gaussian elimination. In any case, a certain version of the matrix that has the maximum number of components eliminated is said to be the Row Reduced Echelon Form (RREF).

**Example 12** (Redundant equations)

$$\left. \begin{array}{ccccc} x & + & y & = & 2 \\ 2x & + & 2y & = & 4 \end{array} \right\} \Leftrightarrow \left( \begin{array}{cc|c} 1 & 1 & 2 \\ 2 & 2 & 4 \end{array} \right) \sim \left( \begin{array}{cc|c} 1 & 1 & 2 \\ 0 & 0 & 0 \end{array} \right) \Leftrightarrow \left\{ \begin{array}{ccccc} x & - & y & = & 2 \\ 0 & - & 0 & = & 0 \end{array} \right.$$

This example demonstrates if one equation is a multiple of the other the identity matrix can not be a reached. This is because the first step in elimination will make the second row a row of zeros. Notice that solutions still exists $(1,1)$ is a solution. The last augmented matrix here is in RREF; no more than two components can be eliminated.

**Example 13** (Inconsistent equations)

$$\left. \begin{array}{ccccc} x & + & y & = & 2 \\ 2x & + & 2y & = & 5 \end{array} \right\} \Leftrightarrow \left( \begin{array}{cc|c} 1 & 1 & 2 \\ 2 & 2 & 5 \end{array} \right) \sim \left( \begin{array}{cc|c} 1 & 1 & 2 \\ 0 & 0 & 1 \end{array} \right) \Leftrightarrow \left\{ \begin{array}{ccccc} x & - & y & = & 2 \\ 0 & - & 0 & = & 1 \end{array} \right.$$

This system of equation has a solution if there exists two numbers $x$, and $y$ such that $0 + 0 = 1$. That is a tricky way of saying there are no solutions. The last form of the augmented matrix here is the RREF.

**Example 14** (Silly order of equations)
A robot might make this mistake:

$$\left. \begin{array}{rcrcr} 0x & + & y & = & -2 \\ x & + & y & = & 7 \end{array} \right\} \Leftrightarrow \left( \begin{array}{cc|c} 0 & 1 & -2 \\ 1 & 1 & 7 \end{array} \right) \sim \cdots,$$

and then give up because the the upper left slot can not function as a pivot since the 0 that lives there can not be used to eliminate the zero below it. Of course, the right thing to do is to change the order of the equations before starting

$$\left. \begin{array}{rcrcr} x & + & y & = & 7 \\ 0x & + & y & = & -2 \end{array} \right\} \Leftrightarrow \left( \begin{array}{cc|c} 1 & 1 & 7 \\ 0 & 1 & -2 \end{array} \right) \sim \left( \begin{array}{cc|c} 1 & 0 & 9 \\ 0 & 1 & -2 \end{array} \right) \Leftrightarrow \left\{ \begin{array}{rcrcr} x & + & 0 & = & 9 \\ 0 & + & y & = & -2. \end{array} \right.$$

The third augmented matrix above is the RREF of the first and second. That is to say, you can swap rows on your way to RREF.

For larger systems of equations redundancy and inconsistency are the obstructions to obtaining the identity matrix, and hence to a simple statement of a solution in the form $x = a, y = b, \ldots$ . What can we do to maximally simplify a system of equations in general? We need to perform operations that simplify our system *without changing its solutions*. Because, exchanging the order of equations, multiplying one equation by a *non-zero* constant or adding equations does not change the system's solutions, we are lead to three operations:

- (Row Swap) Exchange any two rows.

- (Scalar Multiplication) Multiply any row by a non-zero constant.

- (Row Sum) Add a multiple of one row to another row.

These are called *Elementary Row Operations*, or EROs for short, and are studied in detail in section 2.3. Suppose now we have a general augmented matrix for which the first entry in the first row does not vanish. Then, using just the three EROs, we could[1] then perform the following.

---

[1]This is a "brute force" algorithm; there will often be more efficient ways to get to RREF.

# Algorithm For Obtaining RREF:

- Make the leftmost nonzero entry in the top row 1 by multiplication.

- Then use that 1 as a pivot to eliminate everything below it.

- Then go to the next row and make the leftmost nonzero entry 1.

- Use that 1 as a pivot to eliminate everything below *and above it!*

- Go to the next row and make the leftmost nonzero entry 1... *etc*

In the case that the first entry of the first row is zero, we may first interchange the first row with another row whose first entry is non-vanishing and then perform the above algorithm. If the entire first column vanishes, we may still apply the algorithm on the remaining columns.

Here is a video (with special effects!) of a hand performing the algorithm by hand. When it is done, you should try doing what it does.

## Beginner Elimination

This algorithm and its variations is known as Gaussian elimination. The endpoint of the algorithm is an augmented matrix of the form

$$\left( \begin{array}{ccccccc|c} 1 & * & 0 & * & 0 & \cdots & 0 & * & b^1 \\ 0 & 0 & 1 & * & 0 & \cdots & 0 & * & b^2 \\ 0 & 0 & 0 & 0 & 1 & \cdots & 0 & * & b^3 \\ \vdots & \vdots & \vdots & & \vdots & & & \vdots & \vdots \\ 0 & 0 & 0 & 0 & 0 & \cdots & 1 & * & b^k \\ 0 & 0 & 0 & 0 & 0 & \cdots & 0 & 0 & b^{k-1} \\ \vdots & \vdots & \vdots & \vdots & \vdots & & & \vdots & \vdots \\ 0 & 0 & 0 & 0 & 0 & \cdots & 0 & 0 & b^r \end{array} \right).$$

This is called *Reduced Row Echelon Form* (RREF). The asterisks denote the possibility of arbitrary numbers (*e.g.*, the second 1 in the top line of example 12).

Learning to perform this algorithm by hand is the first step to learning linear algebra; it will be the primary means of computation for this course. You need to learn it well. So start practicing as soon as you can, and practice often.

# The following properties define RREF:

1. In every row the left most non-zero entry is 1 (and is called a pivot).

2. The pivot of any given row is always to the right of the pivot of the row above it.

3. The pivot is the only non-zero entry in its column.

**Example 15** (Augmented matrix in RREF)

$$\begin{pmatrix} 1 & 0 & 7 & 0 \\ 0 & 1 & 3 & 0 \\ 0 & 0 & 0 & 1 \\ 0 & 0 & 0 & 0 \end{pmatrix}$$

**Example 16** (Augmented matrix NOT in RREF)

$$\begin{pmatrix} 1 & 0 & 3 & 0 \\ 0 & 0 & 2 & 0 \\ 0 & 1 & 0 & 1 \\ 0 & 0 & 0 & 1 \end{pmatrix}$$

Actually, this NON-example breaks all three of the rules!

The reason we need the asterisks in the general form of RREF is that not every column need have a pivot, as demonstrated in examples 12 and 15. Here is an example where multiple columns have no pivot:

**Example 17** (Consecutive columns with no pivot in RREF)

$$\left. \begin{array}{rcrcrcrcr} x & + & y & + & z & + & 0w & = & 2 \\ 2x & + & 2y & + & 2z & + & 2w & = & 4 \end{array} \right\} \Leftrightarrow \left( \begin{array}{cccc|c} 1 & 1 & 1 & 0 & 2 \\ 2 & 2 & 2 & 1 & 4 \end{array} \right) \sim \left( \begin{array}{cccc|c} 1 & 1 & 1 & 0 & 2 \\ 0 & 0 & 0 & 1 & 0 \end{array} \right)$$

$$\Leftrightarrow \left\{ \begin{array}{rcrcl} x & + & y + z & = & 2 \\ & & w & = & 0. \end{array} \right.$$

Note that there was no hope of reaching the identity matrix, because of the shape of the augmented matrix we started with.

With some practice, elimination can go quickly. Here is an expert showing you some tricks. If you can't follow him now then come back when you have some more experience and watch again. You are going to need to get really good at this!

## Advanced Elimination

It is important that you are able to convert RREF back into a system of equations. The first thing you might notice is that if any of the numbers $b^{k+1}, \ldots, b^r$ in 2.1.3 are non-zero then the system of equations is inconsistent and has no solutions. Our next task is to extract all possible solutions from an RREF augmented matrix.

## Solution Sets and RREF

RREF is a maximally simplified version of the original system of equations in the following sense:

- As many coefficients of the variables as possible are 0.

- As many coefficients of the variables as possible are 1.

It is easier to read off solutions from the maximally simplified equations than from the original equations, even when there are infinitely many solutions.

**Example 18** (Standard approach from a system of equations to the solution set)

$$\left.\begin{array}{rcrcrcl} x & - & y & & & - & 5w & = & 1 \\ & & y & & & - & 2w & = & 6 \\ & & & & z & - & 4w & = & 8 \end{array}\right\} \Leftrightarrow \left(\begin{array}{cccc|c} 1 & 1 & 0 & 5 & 1 \\ 0 & 1 & 0 & 2 & 6 \\ 0 & 0 & 1 & 4 & 8 \end{array}\right) \sim \left(\begin{array}{cccc|c} 1 & 0 & 0 & 3 & -5 \\ 0 & 1 & 0 & 2 & 6 \\ 0 & 0 & 1 & 4 & 8 \end{array}\right)$$

$$\Leftrightarrow \left\{\begin{array}{rcrcl} x & & & + & 3w & = & -5 \\ & y & & + & 2w & = & 6 \\ & & z & + & 4w & = & 8 \end{array}\right\} \Leftrightarrow \left\{\begin{array}{rcl} x & = & -5 & - & 3w \\ y & = & 6 & - & 2w \\ z & = & 8 & - & 4w \\ w & = & & & w \end{array}\right\}$$

$$\Leftrightarrow \begin{pmatrix} x \\ y \\ z \\ w \end{pmatrix} = \begin{pmatrix} -5 \\ 6 \\ 8 \\ 0 \end{pmatrix} - w \begin{pmatrix} -3 \\ -2 \\ -4 \\ 1 \end{pmatrix}.$$

There is one solution for each value of $w$, so the solution set is

$$\left\{ \begin{pmatrix} -5 \\ 6 \\ 8 \\ 0 \end{pmatrix} - \alpha \begin{pmatrix} -3 \\ -2 \\ -4 \\ 1 \end{pmatrix} : \alpha \in \mathbb{R} \right\}.$$

Here is a verbal description of the preceeding example of the *standard approach*. We say that $x, y$, and $z$ are *pivot variables* because they appeared with a pivot coefficient in RREF. Since $w$ never appears with a pivot coefficient, it is not a pivot variable. In the second line we put all the pivot variables on one side and all the *non-pivot variables* on the other side and added the trivial equation $w = w$ to obtain a system that allowed us to easily read off solutions.

# The Standard Approach To Solutions Sets

1. Write the augmented matrix.

2. Perform EROs to reach RREF.

3. Express the non-pivot variables in terms of the pivot variables.

There are always exactly enough non-pivot variables to index your solutions. In any approach, the variables which are not expressed in terms of the other variables are called *free variables*. The standard approach is to use the non-pivot variables as free variables.

Non-standard approach: solve for $w$ in terms of $z$ and substitute into the other equations. You now have an expression for each component in terms of $z$. But why pick $z$ instead of $y$ or $x$? (or $x + y$?) The standard approach not only feels natural, but is *canonical*, meaning that everyone will get the same RREF and hence choose the same variables to be free. However, it is important to remember that so long as their *set* of solutions is the same, any two choices of free variables is fine. (You might think of this as the difference between using Google Maps™ or Mapquest™; although their maps may look different, the place (home *sic*) they are describing is the same!)

When you see an RREF augmented matrix with two columns that have no pivot, you know there will be two free variables.

**Example 19** (Standard approach, multiple free variables)

$$\begin{pmatrix} 1 & 0 & 7 & 0 & | & 4 \\ 0 & 1 & 3 & 4 & | & 1 \\ 0 & 0 & 0 & 0 & | & 0 \\ 0 & 0 & 0 & 0 & | & 0 \end{pmatrix} \Leftrightarrow \left\{ \begin{array}{l} x \quad -7z \quad\quad = 4 \\ y - 3z + 4w = 1 \end{array} \right\}$$

$$\Leftrightarrow \left\{ \begin{array}{l} x = 4 - 7z \\ y = 1 - 3z - 4w \\ z = \quad\quad z \\ w = \quad\quad\quad\quad w \end{array} \right\} \Leftrightarrow \begin{pmatrix} x \\ y \\ z \\ w \end{pmatrix} = \begin{pmatrix} 4 \\ 1 \\ 0 \\ 0 \end{pmatrix} - z \begin{pmatrix} -7 \\ -3 \\ 1 \\ 0 \end{pmatrix} - w \begin{pmatrix} 0 \\ -4 \\ 0 \\ 1 \end{pmatrix}$$

so the solution set is

$$\left\{ \begin{pmatrix} 4 \\ 1 \\ 0 \\ 0 \end{pmatrix} - z \begin{pmatrix} -7 \\ -3 \\ 1 \\ 0 \end{pmatrix} - w \begin{pmatrix} 0 \\ -4 \\ 0 \\ 1 \end{pmatrix} : z, w \in \mathbb{R} \right\}.$$

## From RREF to a Solution Set

You can imagine having three, four, or fifty-six non-pivot columns and the same number of free variables indexing your solutions set. In general a solution set to a system of equations with $n$ free variables will be of the form

$$\{P + \mu_1 H_1 + \mu_2 H_2 + \cdots + \mu_n H_n : \mu_1, \ldots, \mu_n \in \mathbb{R}\}.$$

The parts of these solutions play special roles in the associated matrix equation. This will come up again and again long after we complete this discussion of basic calculation methods, so we will use the general language of linear algebra to give names to these parts now.

**Definition:** A **homogeneous solution** to a linear equation $Lx = v$, with $L$ and $v$ known is a vector $H$ such that $LH = 0$ where $0$ is the zero vector.

If you have a particular solution $P$ to a linear equation and add a sum of multiples of homogeneous solutions to it you obtain another particular solution.

## Particular and Homogeneous Solutions

Check now that the parts of the solutions with free variables as coefficients from the previous examples are homogeneous solutions, and that by adding a homogeneous solution to a particular solution one obtains a solution to the matrix equation. This will come up over and over again. As an example without matrices, consider the differential equation $\frac{d^2}{dx^2}f = 3$. A particular solution is $\frac{3}{2}x^2$ while $x$ and $1$ are homogeneous solutions. The solution set is $\{\frac{3}{2}x^2 + ax + c1 \; : \; a, b \in \mathbb{R}\}$. You can imagine similar differential equations with more homogeneous solutions.

You need to become very adept at reading off solutions sets of linear systems from the RREF of their augmented matrix; it is a basic skill for linear algebra, and we will continue using it up to the last page of the book!

## Worked examples of Gaussian elimination

1. State whether the following augmented matrices are in RREF and compute their solution sets.

$$\left(\begin{array}{ccccc|c} 1 & 0 & 0 & 0 & 3 & 1 \\ 0 & 1 & 0 & 0 & 1 & 2 \\ 0 & 0 & 1 & 0 & 1 & 3 \\ 0 & 0 & 0 & 1 & 2 & 0 \end{array}\right),$$

$$\left(\begin{array}{cccccc|c} 1 & 1 & 0 & 1 & 0 & 1 & 0 \\ 0 & 0 & 1 & 2 & 0 & 2 & 0 \\ 0 & 0 & 0 & 0 & 1 & 3 & 0 \\ 0 & 0 & 0 & 0 & 0 & 0 & 0 \end{array}\right),$$

$$\begin{pmatrix} 1 & 1 & 0 & 1 & 0 & 1 & 0 & | & 1 \\ 0 & 0 & 1 & 2 & 0 & 2 & 0 & | & -1 \\ 0 & 0 & 0 & 0 & 1 & 3 & 0 & | & 1 \\ 0 & 0 & 0 & 0 & 0 & 2 & 0 & | & -2 \\ 0 & 0 & 0 & 0 & 0 & 0 & 1 & | & 1 \end{pmatrix}.$$

2. Solve the following linear system:

$$2x_1 + 5x_2 - 8x_3 + 2x_4 + 2x_5 = 0$$
$$6x_1 + 2x_2 - 10x_3 + 6x_4 + 8x_5 = 6$$
$$3x_1 + 6x_2 + 2x_3 + 3x_4 + 5x_5 = 6$$
$$3x_1 + 1x_2 - 5x_3 + 3x_4 + 4x_5 = 3$$
$$6x_1 + 7x_2 - 3x_3 + 6x_4 + 9x_5 = 9$$

Be sure to set your work out carefully with equivalence signs $\sim$ between each step, labeled by the row operations you performed.

3. Check that the following two matrices are row-equivalent:

$$\begin{pmatrix} 1 & 4 & 7 & | & 10 \\ 2 & 9 & 6 & | & 0 \end{pmatrix} \quad \text{and} \quad \begin{pmatrix} 0 & -1 & 8 & | & 20 \\ 4 & 18 & 12 & | & 0 \end{pmatrix}.$$

Now remove the third column from each matrix, and show that the resulting two matrices (shown below) are row-equivalent:

$$\begin{pmatrix} 1 & 4 & | & 10 \\ 2 & 9 & | & 0 \end{pmatrix} \quad \text{and} \quad \begin{pmatrix} 0 & -1 & | & 20 \\ 4 & 18 & | & 0 \end{pmatrix}.$$

Now remove the fourth column from each of the original two matrices, and show that the resulting two matrices, viewed as augmented matrices (shown below) are row-equivalent:

$$\begin{pmatrix} 1 & 4 & | & 7 \\ 2 & 9 & | & 6 \end{pmatrix} \quad \text{and} \quad \begin{pmatrix} 0 & -1 & | & 8 \\ 4 & 18 & | & 12 \end{pmatrix}.$$

Explain why row-equivalence is never affected by removing columns.

4. Check that the system of equations corresponding to the augmented matrix

$$\begin{pmatrix} 1 & 4 & | & 10 \\ 3 & 13 & | & 9 \\ 4 & 17 & | & 20 \end{pmatrix}$$

has no solutions. If you remove one of the rows of this matrix, does the new matrix have any solutions? In general, can row equivalence be affected by removing rows? Explain why or why not.

5. Explain why the linear system has no solutions:

$$\begin{pmatrix} 1 & 0 & 3 & | & 1 \\ 0 & 1 & 2 & | & 4 \\ 0 & 0 & 0 & | & 6 \end{pmatrix}$$

For which values of $k$ does the system below have a solution?

$$\begin{array}{rcrcrcr} x & - & 3y & & & = & 6 \\ x & & & + & 3z & = & -3 \\ 2x & + & ky & + & (3-k)z & = & 1 \end{array}$$

## Hint

6. Show that the RREF of a matrix is unique. (Hint: Consider what happens if the same augmented matrix had two different RREFs. Try to see what happens if you removed columns from these two RREF augmented matrices.)

7. Another method for solving linear systems is to use row operations to bring the augmented matrix to Row Echelon Form (REF as opposed to RREF). In REF, the pivots are not necessarily set to one, and we only require that all entries left of the pivots are zero, not necessarily entries above a pivot. Provide a counterexample to show that row echelon form is not unique.

Once a system is in row echelon form, it can be solved by "back substitution." Write the following row echelon matrix as a system of equations, then solve the system using back-substitution.

$$\begin{pmatrix} 2 & 3 & 1 & | & 6 \\ 0 & 1 & 1 & | & 2 \\ 0 & 0 & 3 & | & 3 \end{pmatrix}$$

8. Show that this pair of augmented matrices are row equivalent, assuming $ad - bc \neq 0$:

$$\left( \begin{array}{cc|c} a & b & e \\ c & d & f \end{array} \right) \sim \left( \begin{array}{cc|c} 1 & 0 & \frac{de-bf}{ad-bc} \\ 0 & 1 & \frac{af-ce}{ad-bc} \end{array} \right)$$

9. Consider the augmented matrix:

$$\left( \begin{array}{cc|c} 2 & -1 & 3 \\ -6 & 3 & 1 \end{array} \right) .$$

Give a *geometric* reason why the associated system of equations has no solution. (Hint, plot the three vectors given by the columns of this augmented matrix in the plane.) Given a general augmented matrix

$$\left( \begin{array}{cc|c} a & b & e \\ c & d & f \end{array} \right) ,$$

can you find a condition on the numbers $a, b, c$ and $d$ that corresponds to the geometric condition you found?

10. A relation $\sim$ on a set of objects $U$ is an *equivalence relation* if the following three properties are satisfied:

   - Reflexive: For any $x \in U$, we have $x \sim x$.
   - Symmetric: For any $x, y \in U$, if $x \sim y$ then $y \sim x$.
   - Transitive: For any $x, y$ and $z \in U$, if $x \sim y$ and $y \sim z$ then $x \sim z$.

   Show that row equivalence of matrices is an example of an equivalence relation.

11. Equivalence of augmented matrices does not come from equality of their solution sets. Rather, we define two matrices to be equivalent if one can be obtained from the other by elementary row operations.

   Find a pair of augmented matrices that are not row equivalent but do have the same solution set.

# Elementary Row Operations

Elementary row operations are systems of linear equations relating the old and new rows in Gaussian elimination:

**Example 20** (Keeping track of EROs with equations between rows)
We refer to the new $k$th row as $R'_k$ and the old $k$th row as $R_k$.

$$\begin{pmatrix} 0 & 1 & 1 & | & 7 \\ 2 & 0 & 0 & | & 4 \\ 0 & 0 & 1 & | & 4 \end{pmatrix} \quad \begin{matrix} R'_1 = 0R_1 + R_2 - 0R_3 \\ R'_2 = R_1 + 0R_2 - 0R_3 \\ R'_3 = 0R_1 + 0R_2 - R_3 \\ \sim \end{matrix} \quad \begin{pmatrix} 2 & 0 & 0 & | & 4 \\ 0 & 1 & 1 & | & 7 \\ 0 & 0 & 1 & | & 4 \end{pmatrix} \quad \begin{vmatrix} \\ \\ \\ \end{vmatrix} \quad \begin{pmatrix} R'_1 \\ R'_2 \\ R'_3 \end{pmatrix} = \begin{pmatrix} 0 & 1 & 0 \\ 1 & 0 & 0 \\ 0 & 0 & 1 \end{pmatrix} \begin{pmatrix} R_1 \\ R_2 \\ R_3 \end{pmatrix}$$

$$\begin{matrix} R'_1 = \frac{1}{2}R_1 + 0R_2 - 0R_3 \\ R'_2 = 0R_1 + R_2 - 0R_3 \\ R'_3 = 0R_1 + 0R_2 - R_3 \\ \sim \end{matrix} \quad \begin{pmatrix} 1 & 0 & 0 & | & 2 \\ 0 & 1 & 1 & | & 7 \\ 0 & 0 & 1 & | & 4 \end{pmatrix} \quad \begin{pmatrix} R'_1 \\ R'_2 \\ R'_3 \end{pmatrix} = \begin{pmatrix} \frac{1}{2} & 0 & 0 \\ 0 & 1 & 0 \\ 0 & 0 & 1 \end{pmatrix} \begin{pmatrix} R_1 \\ R_2 \\ R_3 \end{pmatrix}$$

$$\begin{matrix} R'_1 = R_1 + 0R_2 - 0R_3 \\ R'_2 = 0R_1 + R_2 - R_3 \\ R'_3 = 0R_1 + 0R_2 - R_3 \\ \sim \end{matrix} \quad \begin{pmatrix} 1 & 0 & 0 & | & 2 \\ 0 & 1 & 0 & | & 3 \\ 0 & 0 & 1 & | & 4 \end{pmatrix} \quad \begin{pmatrix} R'_1 \\ R'_2 \\ R'_3 \end{pmatrix} = \begin{pmatrix} 1 & 0 & 0 \\ 0 & 1 & -1 \\ 0 & 0 & 1 \end{pmatrix} \begin{pmatrix} R_1 \\ R_2 \\ R_3 \end{pmatrix}$$

On the right, we have listed the relations between old and new rows in matrix notation.

## EROs and Matrices

Interestingly, the matrix that describes the relationship between old and new rows performs the corresponding ERO on the augmented matrix.

**Example 21** (Performing EROs with Matrices)

$$\begin{pmatrix} 0 & 1 & 0 \\ 1 & 0 & 0 \\ 0 & 0 & 1 \end{pmatrix} \left(\begin{array}{ccc|c} 0 & 1 & 1 & 7 \\ 2 & 0 & 0 & 4 \\ 0 & 0 & 1 & 4 \end{array}\right) = \left(\begin{array}{ccc|c} 2 & 0 & 0 & 4 \\ 0 & 1 & 1 & 7 \\ 0 & 0 & 1 & 4 \end{array}\right)$$

$$\wr$$

$$\begin{pmatrix} \frac{1}{2} & 0 & 0 \\ 0 & 1 & 0 \\ 0 & 0 & 1 \end{pmatrix} \left(\begin{array}{ccc|c} 2 & 0 & 0 & 4 \\ 0 & 1 & 1 & 7 \\ 0 & 0 & 1 & 4 \end{array}\right) = \left(\begin{array}{ccc|c} 1 & 0 & 0 & 2 \\ 0 & 1 & 1 & 7 \\ 0 & 0 & 1 & 4 \end{array}\right)$$

$$\wr$$

$$\begin{pmatrix} 1 & 0 & 0 \\ 0 & 1 & -1 \\ 0 & 0 & 1 \end{pmatrix} \left(\begin{array}{ccc|c} 1 & 0 & 0 & 2 \\ 0 & 1 & 1 & 7 \\ 0 & 0 & 1 & 4 \end{array}\right) = \left(\begin{array}{ccc|c} 1 & 0 & 0 & 2 \\ 0 & 1 & 0 & 3 \\ 0 & 0 & 1 & 4 \end{array}\right)$$

Here we have multiplied the augmented matrix with the matrices that acted on rows listed on the right of example 20.

Realizing EROs as matrices allows us to give a concrete notion of "dividing by a matrix"; we can now perform manipulations on both sides of an equation in a familiar way:

**Example 22** (Undoing $A$ in $Ax = b$ slowly, for $A = 6 = 3 \cdot 2$)

$$\begin{aligned} 6x &= 12 \\ \Leftrightarrow \quad 3^{-1}6x &= 3^{-1}12 \\ \Leftrightarrow \quad 2x &= 4 \\ \Leftrightarrow \quad 2^{-1}2x &= 2^{-1}4 \\ \Leftrightarrow \quad 1x &= 2 \end{aligned}$$

The matrices corresponding to EROs undo a matrix step by step.

**Example 23** (Undoing $A$ in $Ax = b$ slowly, for $A = M = ...$)

$$\begin{pmatrix} 0 & 1 & 1 \\ 2 & 0 & 0 \\ 0 & 0 & 1 \end{pmatrix} \begin{pmatrix} x \\ y \\ z \end{pmatrix} = \begin{pmatrix} 7 \\ 4 \\ 4 \end{pmatrix}$$

$$\Leftrightarrow \begin{pmatrix} 0 & 1 & 0 \\ 1 & 0 & 0 \\ 0 & 0 & 1 \end{pmatrix} \begin{pmatrix} 0 & 1 & 1 \\ 2 & 0 & 0 \\ 0 & 0 & 1 \end{pmatrix} \begin{pmatrix} x \\ y \\ z \end{pmatrix} = \begin{pmatrix} 0 & 1 & 0 \\ 1 & 0 & 0 \\ 0 & 0 & 1 \end{pmatrix} \begin{pmatrix} 7 \\ 4 \\ 4 \end{pmatrix}$$

$$\Leftrightarrow \begin{pmatrix} 2 & 0 & 0 \\ 0 & 1 & 1 \\ 0 & 0 & 1 \end{pmatrix} \begin{pmatrix} x \\ y \\ z \end{pmatrix} = \begin{pmatrix} 4 \\ 7 \\ 4 \end{pmatrix}$$

$$\Leftrightarrow \begin{pmatrix} \frac{1}{2} & 0 & 0 \\ 0 & 1 & 0 \\ 0 & 0 & 1 \end{pmatrix} \begin{pmatrix} 2 & 0 & 0 \\ 0 & 1 & 1 \\ 0 & 0 & 1 \end{pmatrix} \begin{pmatrix} x \\ y \\ z \end{pmatrix} = \begin{pmatrix} \frac{1}{2} & 0 & 0 \\ 0 & 1 & 0 \\ 0 & 0 & 1 \end{pmatrix} \begin{pmatrix} 4 \\ 7 \\ 4 \end{pmatrix}$$

$$\Leftrightarrow \begin{pmatrix} 1 & 0 & 0 \\ 0 & 1 & 1 \\ 0 & 0 & 1 \end{pmatrix} \begin{pmatrix} x \\ y \\ z \end{pmatrix} = \begin{pmatrix} 2 \\ 7 \\ 4 \end{pmatrix}$$

$$\Leftrightarrow \begin{pmatrix} 1 & 0 & 0 \\ 0 & 1 & -1 \\ 0 & 0 & 1 \end{pmatrix} \begin{pmatrix} 1 & 0 & 0 \\ 0 & 1 & 1 \\ 0 & 0 & 1 \end{pmatrix} \begin{pmatrix} x \\ y \\ z \end{pmatrix} = \begin{pmatrix} 1 & 0 & 0 \\ 0 & 1 & -1 \\ 0 & 0 & 1 \end{pmatrix} \begin{pmatrix} 2 \\ 7 \\ 4 \end{pmatrix}$$

$$\Leftrightarrow \begin{pmatrix} 1 & 0 & 0 \\ 0 & 1 & 0 \\ 0 & 0 & 1 \end{pmatrix} \begin{pmatrix} x \\ y \\ z \end{pmatrix} = \begin{pmatrix} 2 \\ 3 \\ 4 \end{pmatrix}.$$

This is another way of thinking about Gaussian elimination which feels more like elementary algebra in the sense that you "do something to both sides of an equation" until you have a solution.

## Recording EROs in $(M|I)$

Just as we put together $3^{-1}2^{-1} = 6^{-1}$ to get a single thing to apply to both sides of $6x = 12$ to undo 6, we should put together multiple EROs to get a single thing that undoes our matrix. To do this, augment by the identity matrix (not just a single column) and then perform Gaussian elimination. There is no need to write the EROs as systems of equations or as matrices while doing this.

**Example 24** (Collecting EROs that undo a matrix)

$$\left(\begin{array}{ccc|ccc} 0 & 1 & 1 & 1 & 0 & 0 \\ 2 & 0 & 0 & 0 & 1 & 0 \\ 0 & 0 & 1 & 0 & 0 & 1 \end{array}\right) \sim \left(\begin{array}{ccc|ccc} 2 & 0 & 0 & 0 & 1 & 0 \\ 0 & 1 & 1 & 1 & 0 & 0 \\ 0 & 0 & 1 & 0 & 0 & 1 \end{array}\right)$$

$$\sim \left(\begin{array}{ccc|ccc} 1 & 0 & 0 & 0 & \frac{1}{2} & 0 \\ 0 & 1 & 1 & 1 & 0 & 0 \\ 0 & 0 & 1 & 0 & 0 & 1 \end{array}\right) \sim \left(\begin{array}{ccc|ccc} 1 & 0 & 0 & 0 & \frac{1}{2} & 0 \\ 0 & 1 & 0 & 1 & 0 & -1 \\ 0 & 0 & 1 & 0 & 0 & 1 \end{array}\right).$$

As we changed the left side from the matrix $M$ to the identity matrix, the right side changed from the identity matrix to the matrix which undoes $M$.

**Example 25** (Checking that one matrix undoes another)

$$\left(\begin{array}{ccc} 0 & \frac{1}{2} & 0 \\ 1 & 0 & -1 \\ 0 & 0 & 1 \end{array}\right) \left(\begin{array}{ccc} 0 & 1 & 1 \\ 2 & 0 & 0 \\ 0 & 0 & 1 \end{array}\right) = \left(\begin{array}{ccc} 1 & 0 & 0 \\ 0 & 1 & 0 \\ 0 & 0 & 1 \end{array}\right).$$

If the matrices are composed in the opposite order, the result is the same.

$$\left(\begin{array}{ccc} 0 & 1 & 1 \\ 2 & 0 & 0 \\ 0 & 0 & 1 \end{array}\right) \left(\begin{array}{ccc} 0 & \frac{1}{2} & 0 \\ 1 & 0 & -1 \\ 0 & 0 & 1 \end{array}\right) = \left(\begin{array}{ccc} 1 & 0 & 0 \\ 0 & 1 & 0 \\ 0 & 0 & 1 \end{array}\right).$$

Whenever the product of two matrices $MN = I$, we say that $N$ is the inverse of $M$ or $N = M^{-1}$. Conversely $M$ is the inverse of $N$; $M = N^{-1}$.

In abstract generality, let $M$ be some matrix and, as always, let $I$ stand for the identity matrix. Imagine the process of performing elementary row operations to bring $M$ to the identity matrix:

$$(M|I) \sim (E_1 M|E_1) \sim (E_2 E_1 M|E_2 E_1) \sim \cdots \sim (I|\cdots E_2 E_1).$$

The ellipses "$\cdots$" stand for additional EROs. The result is a product of matrices that form a matrix which undoes $M$

$$\cdots E_2 E_1 M = I.$$

This is only true if the RREF of $M$ is the identity matrix.

**Definition:** A matrix $M$ is **invertible** if its RREF is an identity matrix.

# How to find $M^{-1}$
- $(M|I) \sim (I|M^{-1})$

Much use is made of the fact that invertible matrices can be undone with EROs. To begin with, since each elementary row operation has an inverse,

$$M = E_1^{-1} E_2^{-1} \cdots ,$$

while the inverse of $M$ is

$$M^{-1} = \cdots E_2 E_1 .$$

This is symbolically verified by

$$M^{-1} M = \cdots E_2 E_1 \, E_1^{-1} E_2^{-1} \cdots = \cdots E_2 \, E_2^{-1} \cdots = \cdots = I .$$

Thus, if $M$ is invertible, then $M$ can be expressed as the product of EROs. (The same is true for its inverse.) This has the feel of the fundamental theorem of arithmetic (integers can be expressed as the product of primes) or the fundamental theorem of algebra (polynomials can be expressed as the product of |complex| first order polynomials); EROs are building blocks of invertible matrices.

## The Three Elementary Matrices

We now work toward concrete examples and applications. It is surprisingly easy to translate between EROs and matrices that perform EROs. The matrices corresponding to these kinds are close in form to the identity matrix:

- Row Swap: Identity matrix with two rows swapped.

- Scalar Multiplication: Identity matrix with one diagonal entry not 1.

- Row Sum: The identity matrix with one off-diagonal entry not 0.

**Example 26** (Correspondences between EROs and their matrices)

- The row swap matrix that swaps the 2nd and 4th row is the identity matrix with the 2nd and 4th row swapped:

$$\begin{pmatrix} 1 & 0 & 0 & 0 & 0 \\ 0 & 0 & 0 & 1 & 0 \\ 0 & 0 & 1 & 0 & 0 \\ 0 & 1 & 0 & 0 & 0 \\ 0 & 0 & 0 & 0 & 1 \end{pmatrix} .$$

- The scalar multiplication matrix that replaces the 3rd row with 7 times the 3rd row is the identity matrix with 7 in the 3rd row instead of 1:

$$\begin{pmatrix} 1 & 0 & 0 & 0 \\ 0 & 1 & 0 & 0 \\ 0 & 0 & 7 & 0 \\ 0 & 0 & 0 & 1 \end{pmatrix}.$$

- The row sum matrix that replaces the 4th row with the 4th row plus 9 times the 2nd row is the identity matrix with a 9 in the 4th row, 2nd column:

$$\begin{pmatrix} 1 & 0 & 0 & 0 & 0 & 0 & 0 \\ 0 & 1 & 0 & 0 & 0 & 0 & 0 \\ 0 & 0 & 1 & 0 & 0 & 0 & 0 \\ 0 & 9 & 0 & 1 & 0 & 0 & 0 \\ 0 & 0 & 0 & 0 & 1 & 0 & 0 \\ 0 & 0 & 0 & 0 & 0 & 1 & 0 \\ 0 & 0 & 0 & 0 & 0 & 0 & 1 \end{pmatrix}.$$

We can write an explicit factorization of a matrix into EROs by keeping track of the EROs used in getting to RREF.

**Example 27** (Express $M$ from Example 24 as a product of EROs)
Note that in the previous example one of each of the kinds of EROs is used, in the order just given. Elimination looked like

$$M = \begin{pmatrix} 0 & 1 & 1 \\ 2 & 0 & 0 \\ 0 & 0 & 1 \end{pmatrix} \overset{E_1}{\sim} \begin{pmatrix} 2 & 0 & 0 \\ 0 & 1 & 1 \\ 0 & 0 & 1 \end{pmatrix} \overset{E_2}{\sim} \begin{pmatrix} 1 & 0 & 0 \\ 0 & 1 & 1 \\ 0 & 0 & 1 \end{pmatrix} \overset{E_3}{\sim} \begin{pmatrix} 1 & 0 & 0 \\ 0 & 1 & 0 \\ 0 & 0 & 1 \end{pmatrix} = I,$$

where the EROs matrices are

$$E_1 = \begin{pmatrix} 0 & 1 & 0 \\ 1 & 0 & 0 \\ 0 & 0 & 1 \end{pmatrix}, \quad E_2 = \begin{pmatrix} \frac{1}{2} & 0 & 0 \\ 0 & 1 & 0 \\ 0 & 0 & 1 \end{pmatrix}, \quad E_3 = \begin{pmatrix} 1 & 0 & 0 \\ 0 & 1 & -1 \\ 0 & 0 & 1 \end{pmatrix}.$$

The inverse of the ERO matrices (corresponding to the description of the reverse row maniplulations)

$$E_1^{-1} = \begin{pmatrix} 0 & 1 & 0 \\ 1 & 0 & 0 \\ 0 & 0 & 1 \end{pmatrix}, \quad E_2^{-1} = \begin{pmatrix} 2 & 0 & 0 \\ 0 & 1 & 0 \\ 0 & 0 & 1 \end{pmatrix}, \quad E_3^{-1} = \begin{pmatrix} 1 & 0 & 0 \\ 0 & 1 & 1 \\ 0 & 0 & 1 \end{pmatrix}.$$

Multiplying these gives

$$
E_1^{-1}E_2^{-1}E_3^{-1} = \begin{pmatrix} 0 & 1 & 0 \\ 1 & 0 & 0 \\ 0 & 0 & 1 \end{pmatrix} \begin{pmatrix} 2 & 0 & 0 \\ 0 & 1 & 0 \\ 0 & 0 & 1 \end{pmatrix} \begin{pmatrix} 1 & 0 & 0 \\ 0 & 1 & 1 \\ 0 & 0 & 1 \end{pmatrix}
$$

$$
= \begin{pmatrix} 0 & 1 & 0 \\ 1 & 0 & 0 \\ 0 & 0 & 1 \end{pmatrix} \begin{pmatrix} 2 & 0 & 0 \\ 0 & 1 & 1 \\ 0 & 0 & 1 \end{pmatrix} = \begin{pmatrix} 0 & 1 & 1 \\ 2 & 0 & 0 \\ 0 & 0 & 1 \end{pmatrix} = M.
$$

## $LU$, $LDU$, and $LDPU$ Factorizations

The process of elimination can be stopped halfway to obtain decompositions frequently used in large computations in sciences and engineering. The first half of the elimination process is to eliminate entries below the diagonal leaving a matrix which is called *upper triangular*. The elementary matrices which perform this part of the elimination are *lower triangular*, as are their inverses. But putting together the upper triangular and lower triangular parts one obtains the so-called $LU$ factorization.

**Example 28** ($LU$ factorization)

$$
M = \begin{pmatrix} 2 & 0 & -3 & 1 \\ 0 & 1 & 2 & 2 \\ -4 & 0 & 9 & 2 \\ 0 & -1 & 1 & -1 \end{pmatrix} \underset{\sim}{E_1} \begin{pmatrix} 2 & 0 & -3 & 1 \\ 0 & 1 & 2 & 2 \\ 0 & 0 & 3 & 4 \\ 0 & -1 & 1 & -1 \end{pmatrix}
$$

$$
\underset{\sim}{E_2} \begin{pmatrix} 2 & 0 & -3 & 1 \\ 0 & 1 & 2 & 2 \\ 0 & 0 & 3 & 4 \\ 0 & 0 & 3 & 1 \end{pmatrix} \underset{\sim}{E_3} \begin{pmatrix} 2 & 0 & -3 & 1 \\ 0 & 1 & 2 & 2 \\ 0 & 0 & 3 & 4 \\ 0 & 0 & 0 & -3 \end{pmatrix} := U,
$$

where the EROs and their inverses are

$$
E_1 = \begin{pmatrix} 1 & 0 & 0 & 0 \\ 0 & 1 & 0 & 0 \\ 2 & 0 & 1 & 0 \\ 0 & 0 & 0 & 1 \end{pmatrix}, \quad E_2 = \begin{pmatrix} 1 & 0 & 0 & 0 \\ 0 & 1 & 0 & 0 \\ 0 & 0 & 1 & 0 \\ 0 & 1 & 0 & 1 \end{pmatrix}, \quad E_3 = \begin{pmatrix} 1 & 0 & 0 & 0 \\ 0 & 1 & 0 & 0 \\ 0 & 0 & 1 & 0 \\ 0 & 0 & -1 & 1 \end{pmatrix}
$$

$$
E_1^{-1} = \begin{pmatrix} 1 & 0 & 0 & 0 \\ 0 & 1 & 0 & 0 \\ -2 & 0 & 1 & 0 \\ 0 & 0 & 0 & 1 \end{pmatrix}, \quad E_2^{-1} = \begin{pmatrix} 1 & 0 & 0 & 0 \\ 0 & 1 & 0 & 0 \\ 0 & 0 & 1 & 0 \\ 0 & -1 & 0 & 1 \end{pmatrix}, \quad E_3^{-1} = \begin{pmatrix} 1 & 0 & 0 & 0 \\ 0 & 1 & 0 & 0 \\ 0 & 0 & 1 & 0 \\ 0 & 0 & 1 & 1 \end{pmatrix}.
$$

# Elementary Row Operations

Applying inverse elementary matrices to both sides of the equality $U = E_3 E_2 E_1 M$ gives $M = E_1^{-1} E_2^{-1} E_3^{-1} U$ or

$$\begin{pmatrix} 2 & 0 & -3 & 1 \\ 0 & 1 & 2 & 2 \\ -4 & 0 & 9 & 2 \\ 0 & -1 & 1 & -1 \end{pmatrix} = \begin{pmatrix} 1 & 0 & 0 & 0 \\ 0 & 1 & 0 & 0 \\ -2 & 0 & 1 & 0 \\ 0 & 0 & 0 & 1 \end{pmatrix} \begin{pmatrix} 1 & 0 & 0 & 0 \\ 0 & 1 & 0 & 0 \\ 0 & 0 & 1 & 0 \\ 0 & -1 & 0 & 1 \end{pmatrix} \begin{pmatrix} 1 & 0 & 0 & 0 \\ 0 & 1 & 0 & 0 \\ 0 & 0 & 1 & 0 \\ 0 & 0 & 1 & 1 \end{pmatrix} \begin{pmatrix} 2 & 0 & -3 & 1 \\ 0 & 1 & 2 & 2 \\ 0 & 0 & 3 & 4 \\ 0 & 0 & 0 & -3 \end{pmatrix}$$

$$= \begin{pmatrix} 1 & 0 & 0 & 0 \\ 0 & 1 & 0 & 0 \\ -2 & 0 & 1 & 0 \\ 0 & 0 & 0 & 1 \end{pmatrix} \begin{pmatrix} 1 & 0 & 0 & 0 \\ 0 & 1 & 0 & 0 \\ 0 & 0 & 1 & 0 \\ 0 & -1 & 1 & 1 \end{pmatrix} \begin{pmatrix} 2 & 0 & -3 & 1 \\ 0 & 1 & 2 & 2 \\ 0 & 0 & 3 & 4 \\ 0 & 0 & 0 & -3 \end{pmatrix}$$

$$= \begin{pmatrix} 1 & 0 & 0 & 0 \\ 0 & 1 & 0 & 0 \\ -2 & 0 & 1 & 0 \\ 0 & -1 & 1 & 1 \end{pmatrix} \begin{pmatrix} 2 & 0 & -3 & 1 \\ 0 & 1 & 2 & 2 \\ 0 & 0 & 3 & 4 \\ 0 & 0 & 0 & -3 \end{pmatrix} .$$

This is a lower triangular matrix times an upper triangular matrix.

What if we stop at a different point in elimination? We could multiply rows so that the entries in the diagonal are 1 next. Note that the EROs that do this are diagonal. This gives a slightly different factorization.

**Example 29** (*LDU* factorization building from previous example)

$$
M = \begin{pmatrix} 2 & 0 & -3 & 1 \\ 0 & 1 & 2 & 2 \\ -4 & 0 & 9 & 2 \\ 0 & -1 & 1 & -1 \end{pmatrix}
\underset{\sim}{E_3 E_2 E_1}
\begin{pmatrix} 2 & 0 & -3 & 1 \\ 0 & 1 & 2 & 2 \\ 0 & 0 & 3 & 4 \\ 0 & 0 & 0 & -3 \end{pmatrix}
\underset{\sim}{E_4}
\begin{pmatrix} 1 & 0 & \frac{-3}{2} & \frac{1}{2} \\ 0 & 1 & 2 & 2 \\ 0 & 0 & 3 & 4 \\ 0 & 0 & 0 & -3 \end{pmatrix}
$$

$$
\underset{\sim}{E_5}
\begin{pmatrix} 1 & 0 & \frac{-3}{2} & \frac{1}{2} \\ 0 & 1 & 2 & 2 \\ 0 & 0 & 1 & \frac{4}{3} \\ 0 & 0 & 0 & -3 \end{pmatrix}
\underset{\sim}{E_6}
\begin{pmatrix} 1 & 0 & \frac{-3}{2} & \frac{1}{2} \\ 0 & 1 & 2 & 2 \\ 0 & 0 & 1 & \frac{4}{3} \\ 0 & 0 & 0 & 1 \end{pmatrix} =: U
$$

The corresponding elementary matrices are

$$
E_4 = \begin{pmatrix} \frac{1}{2} & 0 & 0 & 0 \\ 0 & 1 & 0 & 0 \\ 0 & 0 & 1 & 0 \\ 0 & 0 & 0 & 1 \end{pmatrix}, \quad
E_5 = \begin{pmatrix} 1 & 0 & 0 & 0 \\ 0 & 1 & 0 & 0 \\ 0 & 0 & \frac{1}{3} & 0 \\ 0 & 0 & 0 & 1 \end{pmatrix}, \quad
E_6 = \begin{pmatrix} 1 & 0 & 0 & 0 \\ 0 & 1 & 0 & 0 \\ 0 & 0 & 1 & 0 \\ 0 & 0 & 0 & -\frac{1}{3} \end{pmatrix},
$$

$$
E_4^{-1} = \begin{pmatrix} 2 & 0 & 0 & 0 \\ 0 & 1 & 0 & 0 \\ 0 & 0 & 1 & 0 \\ 0 & 0 & 0 & 1 \end{pmatrix}, \quad
E_5^{-1} = \begin{pmatrix} 1 & 0 & 0 & 0 \\ 0 & 1 & 0 & 0 \\ 0 & 0 & 3 & 0 \\ 0 & 0 & 0 & 1 \end{pmatrix}, \quad
E_6^{-1} = \begin{pmatrix} 1 & 0 & 0 & 0 \\ 0 & 1 & 0 & 0 \\ 0 & 0 & 1 & 0 \\ 0 & 0 & 0 & -3 \end{pmatrix}.
$$

The equation $U = E_6 E_5 E_4 E_3 E_2 E_1 M$ can be rearranged as

$$
M = (E_1^{-1} E_2^{-1} E_3^{-1})(E_4^{-1} E_5^{-1} E_6^{-1})U.
$$

We calculated the product of the first three factors in the previous example; it was named $L$ there, and we will reuse that name here. The product of the next three factors is diagonal and we wil name it $D$. The last factor we named $U$ (the name means something different in this example than the last example.) The *LDU* factorization of our matrix is

$$
\begin{pmatrix} 2 & 0 & -3 & 1 \\ 0 & 1 & 2 & 2 \\ -4 & 0 & 9 & 2 \\ 0 & -1 & 1 & -1 \end{pmatrix} =
\begin{pmatrix} 1 & 0 & 0 & 0 \\ 0 & 1 & 0 & 0 \\ -2 & 0 & 1 & 0 \\ 0 & -1 & 1 & 1 \end{pmatrix}
\begin{pmatrix} 2 & 0 & 0 & 0 \\ 0 & 1 & 0 & 0 \\ 0 & 0 & 3 & 0 \\ 0 & 0 & 1 & -3 \end{pmatrix}
\begin{pmatrix} 1 & 0 & -\frac{3}{2} & \frac{1}{2} \\ 0 & 1 & 2 & 2 \\ 0 & 0 & 1 & \frac{4}{3} \\ 0 & 0 & 0 & 1 \end{pmatrix}.
$$

The $LDU$ factorization of a matrix is a factorization into blocks of EROs of a various types: $L$ is the product of the inverses of EROs which eliminate below the diagonal by row addition, $D$ the product of inverses of EROs which set the diagonal elements to 1 by row multiplication, and $U$ is the product of inverses of EROs which eliminate above the diagonal by row addition.

You may notice that one of the three kinds of row operation is missing from this story. Row exchange may be necessary to obtain RREF. Indeed, so far in this chapter we have been working under the tacit assumption that $M$ can be brought to the identity by just row multiplication and row addition. If row exchange is necessary, the resulting factorization is $LDPU$ where $P$ is the product of inverses of EROs that perform row exchange.

**Example 30** ($LDPU$ factorization, building from previous examples)

$$M = \begin{pmatrix} 0 & 1 & 2 & 2 \\ 2 & 0 & -3 & 1 \\ -4 & 0 & 9 & 2 \\ 0 & -1 & 1 & -1 \end{pmatrix} \underset{\sim}{\overset{E_7}{}} \begin{pmatrix} 2 & 0 & -3 & 1 \\ 0 & 1 & 2 & 2 \\ -4 & 0 & 9 & 2 \\ 0 & -1 & 1 & -1 \end{pmatrix} \underset{\sim}{\overset{E_6 E_5 E_4 E_3 E_2 E_1}{}} L$$

$$E_7 = \begin{pmatrix} 0 & 1 & 0 & 0 \\ 1 & 0 & 0 & 0 \\ 0 & 0 & 1 & 0 \\ 0 & 0 & 0 & 1 \end{pmatrix} = E_7^{-1}$$

$$M = (E_1^{-1} E_2^{-1} E_3^{-1})(E_4^{-1} E_5^{-1} E_6^{-1})(E_7^{-1})U = LDPU$$

$$\begin{pmatrix} 0 & 1 & 2 & 2 \\ 2 & 0 & -3 & 1 \\ -4 & 0 & 9 & 2 \\ 0 & -1 & 1 & -1 \end{pmatrix} = \begin{pmatrix} 1 & 0 & 0 & 0 \\ 0 & 1 & 0 & 0 \\ -2 & 0 & 1 & 0 \\ 0 & -1 & 1 & 1 \end{pmatrix} \begin{pmatrix} 2 & 0 & 0 & 0 \\ 0 & 1 & 0 & 0 \\ 0 & 0 & 3 & 0 \\ 0 & 0 & 1 & -3 \end{pmatrix} \begin{pmatrix} 0 & 1 & 0 & 0 \\ 1 & 0 & 0 & 0 \\ 0 & 0 & 1 & 0 \\ 0 & 0 & 0 & 1 \end{pmatrix} \begin{pmatrix} 1 & 0 & -\frac{3}{2} & \frac{1}{2} \\ 0 & 1 & 2 & 2 \\ 0 & 0 & 1 & \frac{4}{3} \\ 0 & 0 & 0 & 1 \end{pmatrix}$$

1. While performing Gaussian elimination on these augmented matrices write the full system of equations describing the new rows in terms of the old rows above each equivalence symbol as in example 20.

$$\begin{pmatrix} 2 & 2 & | & 10 \\ 1 & 2 & | & 8 \end{pmatrix}, \quad \begin{pmatrix} 1 & 1 & 0 & | & 5 \\ 1 & 1 & -1 & | & 11 \\ -1 & 1 & 1 & | & -5 \end{pmatrix}$$

2. Solve the vector equation by applying ERO matrices to each side of the equation to perform elimination. Show each matrix explicitly as in example 23.

$$\begin{pmatrix} 3 & 6 & 2 \\ 5 & 9 & 4 \\ 2 & 4 & 2 \end{pmatrix} \begin{pmatrix} x \\ y \\ z \end{pmatrix} = \begin{pmatrix} -3 \\ 1 \\ 0 \end{pmatrix}$$

3. Solve this vector equation by finding the inverse of the matrix through $(M|I) \sim (I|M^{-1})$ and then applying $M^{-1}$ to both sides of the equation.

$$\begin{pmatrix} 2 & 1 & 1 \\ 1 & 1 & 1 \\ 1 & 1 & 2 \end{pmatrix} \begin{pmatrix} x \\ y \\ z \end{pmatrix} = \begin{pmatrix} 9 \\ 6 \\ 7 \end{pmatrix}$$

4. Follow the method of examples 28 and 29 to find the $LU$ and $LDU$ factorization of

$$\begin{pmatrix} 3 & 3 & 6 \\ 3 & 5 & 2 \\ 6 & 2 & 5 \end{pmatrix}.$$

5. Multiple matrix equations with the same matrix can be solved simultaneously.

   (a) Solve both systems by performing elimination on just one augmented matrix.

$$\begin{pmatrix} 2 & -1 & -1 \\ -1 & 1 & 1 \\ 1 & -1 & 0 \end{pmatrix} \begin{pmatrix} x \\ y \\ z \end{pmatrix} = \begin{pmatrix} 0 \\ 1 \\ 0 \end{pmatrix}, \quad \begin{pmatrix} 2 & -1 & -1 \\ -1 & 1 & 1 \\ 1 & -1 & 0 \end{pmatrix} \begin{pmatrix} a \\ b \\ c \end{pmatrix} = \begin{pmatrix} 2 \\ 1 \\ 1 \end{pmatrix}$$

    (b) What are the columns of $M^{-1}$ in $(M|I) \sim (I|M^{-1})$?

6. How can you convince your fellow students to never make this mistake?

$$\begin{pmatrix} 1 & 0 & 2 & | & 3 \\ 0 & 1 & 2 & | & 3 \\ 2 & 0 & 1 & | & 4 \end{pmatrix} \begin{matrix} R'_1 = R_1 + R_2 \\ R'_2 = R_1 - R_2 \\ R'_3 = R_1 + 2R_2 \\ \sim \end{matrix} \begin{pmatrix} 1 & 1 & 4 & | & 6 \\ 1 & -1 & 0 & | & 0 \\ 1 & 2 & 6 & | & 9 \end{pmatrix}$$

7. Is $LU$ factorization of a matrix unique? Justify your answer.

∞. If you randomly create a matrix by picking numbers out of the blue, it will probably be difficult to perform elimination or factorization; fractions and large numbers will probably be involved. To invent simple problems it is better to start with a simple answer:

    (a) Start with any augmented matrix in RREF. Perform EROs to make most of the components non-zero. Write the result on a separate piece of paper and give it to your friend. Ask that friend to find RREF of the augmented matrix you gave them. Make sure they get the same augmented matrix you started with.

    (b) Create an upper triangular matrix $U$ and a lower triangular matrix $L$ with only 1s on the diagonal. Give the result to a friend to factor into $LU$ form.

    (c) Do the same with an $LDU$ factorization.

# Solution Sets for Systems of Linear Equations

Algebraic equations problems can have multiple solutions. For example $x(x-1) = 0$ has two solutions: 0 and 1. By contrast, equations of the form $Ax = b$ with $A$ a linear operator (with scalars the real numbers) have the following property:

If $A$ is a linear operator and $b$ is known, then $Ax = b$ has either

1. One solution

2. No solutions

3. Infinitely many solutions

# The Geometry of Solution Sets: Hyperplanes

Consider the following algebra problems and their solutions.

1. $6x = 12$ has one solution: 2.

2a. $0x = 12$ has no solution.

2b. $0x \quad 0$ has infinitely many solutions; its solution set is $\mathbb{R}$.

In each case the linear operator is a $1 \times 1$ matrix. In the first case, the linear operator is invertible. In the other two cases it is not. In the first case, the solution set is a point on the number line, in case 2b the solution set is the whole number line.

Lets examine similar situations with larger matrices: $2 \times 2$ matrices.

1. $\begin{pmatrix} 6 & 0 \\ 0 & 2 \end{pmatrix} \begin{pmatrix} x \\ y \end{pmatrix} = \begin{pmatrix} 12 \\ 6 \end{pmatrix}$ has one solution: $\begin{pmatrix} 2 \\ 3 \end{pmatrix}$.

2a. $\begin{pmatrix} 1 & 3 \\ 0 & 0 \end{pmatrix} \begin{pmatrix} x \\ y \end{pmatrix} = \begin{pmatrix} 4 \\ 1 \end{pmatrix}$ has no solutions.

2bi. $\begin{pmatrix} 1 & 3 \\ 0 & 0 \end{pmatrix} \begin{pmatrix} x \\ y \end{pmatrix} = \begin{pmatrix} 4 \\ 0 \end{pmatrix}$ has solution set $\left\{ \begin{pmatrix} 4 \\ 0 \end{pmatrix} + y \begin{pmatrix} -3 \\ 1 \end{pmatrix} : y \in \mathbb{R} \right\}$.

2bii. $\begin{pmatrix} 0 & 0 \\ 0 & 0 \end{pmatrix} \begin{pmatrix} x \\ y \end{pmatrix} = \begin{pmatrix} 0 \\ 0 \end{pmatrix}$ has solution set $\left\{ \begin{pmatrix} x \\ y \end{pmatrix} : x, y \in \mathbb{R} \right\}$.

Again, in the first case the linear operator is invertible while in the other cases it is not. When a $2 \times 2$ matrix from a matrix equation is not invertible the solution set can be empty, a line in the plane, or the plane itself.

For a system of equations with $r$ equations and $k$ variables, one can have a number of different outcomes. For example, consider the case of $r$ equations in three variables. Each of these equations is the equation of a plane in three-dimensional space. To find solutions to the system of equations, we look for the common intersection of the planes (if an intersection exists). Here we have five different possibilities:

1. **Unique Solution.** The planes have a unique point of intersection.

2a. **No solutions.** Some of the equations are contradictory, so no solutions exist.

2bi. **Line.** The planes intersect in a common line; any point on that line then gives a solution to the system of equations.

2bii. **Plane.** Perhaps you only had one equation to begin with, or else all of the equations coincide geometrically. In this case, you have a plane of solutions, with two free parameters.

## Planes

2biii. **All of $\mathbb{R}^3$.** If you start with no information, then any point in $\mathbb{R}^3$ is a solution. There are three free parameters.

In general, for systems of equations with $k$ unknowns, there are $k + 2$ possible outcomes, corresponding to the possible numbers (*i.e.*, $0, 1, 2, \ldots, k$) of free parameters in the solutions set, plus the possibility of no solutions. These types of solution sets are hyperplanes, generalizations of planes that behave like planes in $\mathbb{R}^3$ in many ways.

## Pictures and Explanation

## Particular Solution + Homogeneous Solutions

Lets look at solution sets again, this time trying to get to their geometric shape. In the standard approach, variables corresponding to columns that do not contain a pivot (after going to reduced row echelon form) are *free*. It is the number of free variables that determines the geometry of the solution set.

**Example 31** (Non-pivot variables determine the gemometry of the solution set)

$$\begin{pmatrix} 1 & 0 & 1 & -1 \\ 0 & 1 & -1 & 1 \\ 0 & 0 & 0 & 0 \end{pmatrix} \begin{pmatrix} x_1 \\ x_2 \\ x_3 \\ x_4 \end{pmatrix} = \begin{pmatrix} 1 \\ 1 \\ 0 \end{pmatrix} \Leftrightarrow \begin{cases} 1x_1 + 0x_2 + 1x_3 - 1x_4 &= 1 \\ 0x_1 + 1x_2 - 1x_3 + 1x_4 &= 1 \\ 0x_1 + 0x_2 + 0x_3 + 0x_4 &= 0 \end{cases}$$

Following the standard approach, express the pivot variables in terms of the non-pivot variables and add "empty equations". Here $x_3$ and $x_4$ are non-pivot variables.

$$\left.\begin{aligned} x_1 &= 1 - x_3 + x_4 \\ x_2 &= 1 - x_3 - x_4 \\ x_3 &= \phantom{1-}x_3 \\ x_4 &= \phantom{1-x_3-}x_4 \end{aligned}\right\} \Leftrightarrow \begin{pmatrix} x_1 \\ x_2 \\ x_3 \\ x_4 \end{pmatrix} = \begin{pmatrix} 1 \\ 1 \\ 0 \\ 0 \end{pmatrix} - x_3 \begin{pmatrix} -1 \\ 1 \\ 1 \\ 0 \end{pmatrix} + x_4 \begin{pmatrix} 1 \\ -1 \\ 0 \\ 1 \end{pmatrix}$$

The preferred way to write a solution set $S$ is with set notation;

$$S = \left\{ \begin{pmatrix} x_1 \\ x_2 \\ x_3 \\ x_4 \end{pmatrix} = \begin{pmatrix} 1 \\ 1 \\ 0 \\ 0 \end{pmatrix} + \mu_1 \begin{pmatrix} -1 \\ 1 \\ 1 \\ 0 \end{pmatrix} + \mu_2 \begin{pmatrix} 1 \\ -1 \\ 0 \\ 1 \end{pmatrix} : \mu_1, \mu_2 \in \mathbb{R} \right\}.$$

Notice that the first two components of the second two terms come from the non-pivot columns. Another way to write the solution set is

$$S = \{ P + \mu_1 H_1 + \mu_2 H_2 : \mu_1, \mu_2 \in \mathbb{R} \} \,,$$

where

$$P = \begin{pmatrix} 1 \\ 1 \\ 0 \\ 0 \end{pmatrix}, \quad H_1 = \begin{pmatrix} -1 \\ 1 \\ 1 \\ 0 \end{pmatrix}, \quad H_2 = \begin{pmatrix} 1 \\ -1 \\ 0 \\ 1 \end{pmatrix}.$$

Here $P$ is a *particular solution* while $H_1$ and $H_2$ are called *homogeneous solutions*. The solution set forms a plane.

## Solutions and Linearity

Motivated by example 31, we say that the matrix equation $MX = V$ has solution set $\{ P + \mu_1 H_1 + \mu_2 H_2 \,|\, \mu_1, \mu_2 \in \mathbb{R} \}$. Recall that matrices are linear operators. Thus

$$M(P + \mu_1 H_1 + \mu_2 H_2) = MP + \mu_1 M H_1 + \mu_2 M H_2 = V \,,$$

for *any* $\mu_1, \mu_2 \in \mathbb{R}$. Choosing $\mu_1 = \mu_2 = 0$, we obtain

$$MP = V \,.$$

This is why $P$ is an example of a *particular solution*.

Setting $\mu_1 = 1, \mu_2 = 0$, and subtracting $MP = V$ we obtain

$$MH_1 = 0.$$

Likewise, setting $\mu_1 = 0, \mu_2 = 1$, we obtain

$$MH_2 = 0.$$

Here $H_1$ and $H_2$ are examples of what are called *homogeneous* solutions to the system. They *do not* solve the original equation $MX = V$, but instead its associated *homogeneous equation* $MY = 0$.

We have just learnt a fundamental lesson of linear algebra: the solution set to $Ax = b$, where $A$ is a linear operator, consists of a particular solution plus homogeneous solutions.

$$\{\text{Solutions}\} = \{\text{Particular solution} + \text{Homogeneous solutions}\}$$

**Example 32** Consider the matrix equation of example 31. It has solution set

$$S = \left\{ \begin{pmatrix} 1 \\ 1 \\ 0 \\ 0 \end{pmatrix} + \mu_1 \begin{pmatrix} -1 \\ 1 \\ 1 \\ 0 \end{pmatrix} + \mu_2 \begin{pmatrix} 1 \\ -1 \\ 0 \\ 1 \end{pmatrix} \, \middle| \, \mu_1, \mu_2 \in \mathbb{R} \right\}.$$

Then $MP = V$ says that $\begin{pmatrix} 1 \\ 1 \\ 0 \\ 0 \end{pmatrix}$ is a solution to the original matrix equation, which is

certainly true, but this is not the only solution.

$MH_1 = 0$ says that $\begin{pmatrix} -1 \\ 1 \\ 1 \\ 0 \end{pmatrix}$ is a solution to the homogeneous equation.

$MH_2 = 0$ says that $\begin{pmatrix} 1 \\ -1 \\ 0 \\ 1 \end{pmatrix}$ is a solution to the homogeneous equation.

Notice how adding any multiple of a homogeneous solution to the particular solution yields another particular solution.

# Review Problems

1. Write down examples of augmented matrices corresponding to each of the five types of solution sets for systems of equations with three unknowns.

2. Invent simple linear system that has multiple solutions. Use the standard approach for solving linear systems and a non-standard approach to obtain different descriptions of the solution set. Is the solution set different with different approaches?

3. Let

$$
M = \begin{pmatrix} a_1^1 & a_2^1 & \cdots & a_k^1 \\ a_1^2 & a_2^2 & \cdots & a_k^2 \\ \vdots & \vdots & & \vdots \\ a_1^r & a_2^r & \cdots & a_k^r \end{pmatrix} \quad \text{and} \quad X = \begin{pmatrix} x^1 \\ x^2 \\ \vdots \\ x^k \end{pmatrix}.
$$

Note: $x^2$ does not denote the square of $x$. Instead $x^1$, $x^2$, $x^3$, etc..., denote different variables; the superscript is an index. Although confusing at first, this notation was invented by Albert Einstein who noticed that quantities like $a_1^2 x^1 + a_2^2 x^2 \cdots + a_k^2 x^k =: \sum_{j=1}^{k} a_j^2 x^j$, can be written unambiguously as $a_j^2 x^j$. This is called Einstein summation notation. The most important thing to remember is that the index $j$ is a dummy variable, so that $a_j^2 x^j \equiv a_i^2 x^i$; this is called "relabeling dummy indices". When dealing with products of sums, you must remember to introduce a new dummy for each term; *i.e.*, $a_i x^i b_i y^i = \sum_i a_i x^i b_i y^i$ does *not* equal $a_i x^i b_j y^j = \left( \sum_i a_i x^i \right) \left( \sum_j b_j y^j \right)$.

Use Einstein summation notation to propose a rule for $MX$ so that $MX = 0$ is equivalent to the linear system

$$
\begin{aligned}
a_1^1 x^1 + a_2^1 x^2 \cdots + a_k^1 x^k &= 0 \\
a_1^2 x^1 + a_2^2 x^2 \cdots + a_k^2 x^k &= 0 \\
\vdots \quad \vdots \qquad \vdots \quad \vdots \quad & \\
a_1^r x^1 + a_2^r x^2 \cdots + a_k^r x^k &= 0
\end{aligned}
$$

Show that your rule for multiplying a matrix by a vector obeys the linearity property.

4. The *standard basis vector* $e_i$ is a column vector with a one in the $i$th row, and zeroes everywhere else. Using the rule for multiplying a matrix times a vector in problem 3, find a simple rule for multiplying $Me_i$, where $M$ is the general matrix defined there.

5. If $A$ is a non-linear operator, can the solutions to $Ax = b$ still be written as "general solution=particular solution + homogeneous solutions"? Provide examples.

6. Find a system of equations whose solution set is the walls of a $1 \times 1 \times 1$ cube. (Hint: You may need to restrict the ranges of the variables; could your equations be linear?)

www.ingramcontent.com/pod-product-compliance
Lightning Source LLC
Chambersburg PA
CBHW080548190526
45169CB00007B/2685